INDUSTRIAL APPLICATIONS OF FUZZY CONTROL

INDUSTRIAL APPLICATIONS OF FUZZY CONTROL

edited by

Michio SUGENO
Department of Systems Science
Graduate School of Science
and Engineering at Nagatsuta
Tokyo Institute of Technology
Yokohama, Japan

1985

NORTH-HOLLAND – AMSTERDAM • NEW YORK • OXFORD

ISBN: 0 444 87829 7

Publishers:

ELSEVIER SCIENCE PUBLISHERS B.V.
P.O. BOX 1991
1000 BZ AMSTERDAM
THE NETHERLANDS

Sole distributors for the U.S.A. and Canada:

ELSEVIER SCIENCE PUBLISHING COMPANY, INC.
52 VANDERBILT AVENUE
NEW YORK, N.Y. 10017
U.S.A.

Library of Congress Cataloging-in-Publication Data
Main entry under title:

Industrial applications of fuzzy control.

Bibliography: p.
1. Automatic control--Addresses, essays, lectures.
2. Fuzzy systems--Addresses, essays, lectures.
I. Sugeno, Michio, 1940- . II. Title: Fuzzy control.
TJ213.7.I47 1985 629.8 85-15908
ISBN 0-444-87829-7 (U.S.)

PRINTED IN THE NETHERLANDS

PREFACE

Many edited volumes on fuzzy sets, fuzzy logic, or fuzzy systems have been published so far. Most of them are the collection of papers on general topics. This volume on "Industrial Applications of Fuzzy Control" is a very special one focussed on the specific field "fuzzy control". It contains only the papers concerned with practical applications of fuzzy control.

Fuzzy control is one of the most promising research fields in fuzzy engineering. Many studies have been devoted to the theory of fuzzy control and its applications since Mamdani's pioneering work. We are now entering a new generation of fuzzy control, where more and more applications to industrial processes are being undertaken.

The editor hopes that this volume of outstanding contributions, including "An Annotated Bibliography of Fuzzy Control", will provide fruitful material to control engineers who are interested in applying fuzzy control to various processes.

May 1985

Michio Sugeno
Editor

ACKNOWLEDGEMENTS

I greatly appreciate the cooperation by all of the contributors to this volume. My special thanks goes to Dr. Richard M. Tong for his help in improving English. My deepest gratitude belongs to Miss Kuniko Miyakawa, who helped me extensively with preparing this volume for publication.

Michio Sugeno
Editor

TABLE OF CONTENTS

Industrial Applications of Fuzzy Control
M. Sugeno (ed.)

AUTOMATIC TRAIN OPERATION SYSTEM BY PREDICTIVE FUZZY CONTROL

Seiji Yasunobu and Shoji Miyamoto

Systems Development Laboratory, Hitachi Ltd.
1099 Ohzenji Asao-ku Kawasaki 215, JAPAN

A predictive fuzzy controller that uses rules based on a skilled human operator's experience has been proposed and applied to Automatic Train Operation (ATO) Systems. The predictive fuzzy controller selects the most likely control command based on a prediction of control results and direct evaluation for the control objectives. The control rules are described as follows; "If (u is C ➔ x is A and y is B) then u is C." The proposed fuzzy controller has been applied to ATO systems which control trains based on the evaluation of safety, riding comfort, traceability of target velocity, accuracy of stop gap, running time and energy consumption.
The simulation results show that the above performance indices of the newly developed Fuzzy ATO are much better than those of conventional PID controlled ATO. Further it is proved by field tests that the Fuzzy ATO can operate trains as skillfully as human experts do.

1. Introduction

In recent years, automatic controllers using a micro-computer instead of a human operator have been developed for plants, transportation systems and so on. In many cases, a computer control gives quick response and accurate control, but inferior quality of control than a skilled human operator. (A fuzzy logic[1] control method which can make up an algorithm from control know-how of a skilled human operator by fuzzy sets, is proposed by Mamdani[2] (1973), and applied to a plant[2], a traffic junction[3], a cement kiln[4], a water treatment[5] and so on[6]. By these controllers, operations considering the desired states of a system are difficult.

To solve this problem, we have proposed a predictive fuzzy controller which predicts the result of each candidate control command and selects the most likely control rule based on a skilled human operator's experience. The predictive fuzzy controller is applied to automatic train operation systems, which are referred to as ATO, by which a train can be started, kept to a limited speed and stopped at a target position of a station.[7,8]

ATOs have been applied to develop new transit systems including subway[9], monorail and new public transit systems like PRT.[10,11] In these systems, trains are controlled by a conventional feedback control method such as a PID control. Therefore, it does not take important performance indices for a train operation such as, safety, riding comfort of passengers, accuracy of the stop gap and energy consumption into consideration.

In this paper, the predictive fuzzy control method and an ATO controller developed based on this method are explained. This controller can realize a skilled human operator's control by evaluating safety, riding comfort, traceability of target velocity, energy saving , running time and the accuracy of a stop gap. By computer simulations, the developed fuzzy ATO controller is compared with the conventional PID ATO controller.

2. Conventional Control and Fuzzy Control

2.1 Conventional control and its bounds

An outline of designs for computer control systems is as follows.

(1) Analysis of an actual system

(2) Decisions of a linearized system model, a desired state and an error criterion function

(3) Syntheses of a controller structure and the controller's parameters

In the above-mentioned design, the control parameters are easy decided by a computer aided design system using a criterion function. But it is difficult to define such a function that sufficiently covers the system's purposes.

The conventional controller is able to give accurate control for systems which are matched to the linearized system models and/or the desired state are constant. However, this method cannot provide adequate control for systems which have time-varying parameters, unknown structures and multi-objects for the control. So, the controller is designed using a complex structured algorithm with the designer's knowledge. But re-constructions of the controller, according to the alternation of a sub-system or change of the system structure, devote a designer's effort every time. Fig.1 illustrates graphically the control sequence.

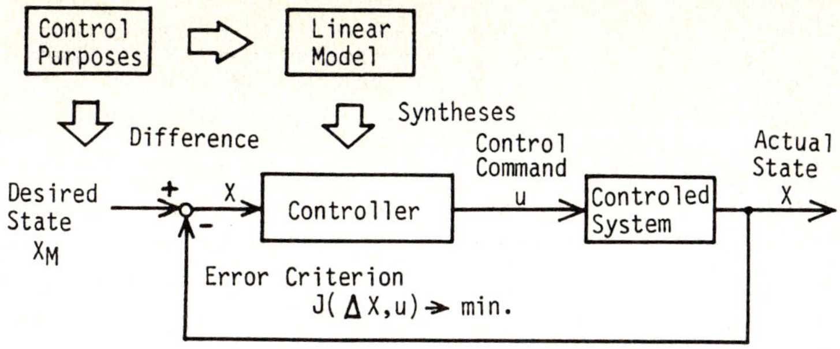

Fig.1 Conventional control sequence

2.2 Application of fuzzy logic to control

2.2.1 States evaluate fuzzy control

The fuzzy control applied to a cement kiln by Holmblad(1982)[4] decides a control action u^* from the set of control rules Ri, which is described as "If x is Ai and Bi, then u is Ui". This fuzzy control evaluates the system state, which is described as "if part: X is A and Y is B", and decides u^* from these values. "If the temperature is high and the pressure is slightly high, then the fuel is decrease." is a typical example of process control rules. Fig.2 illustrates graphically the control sequence.

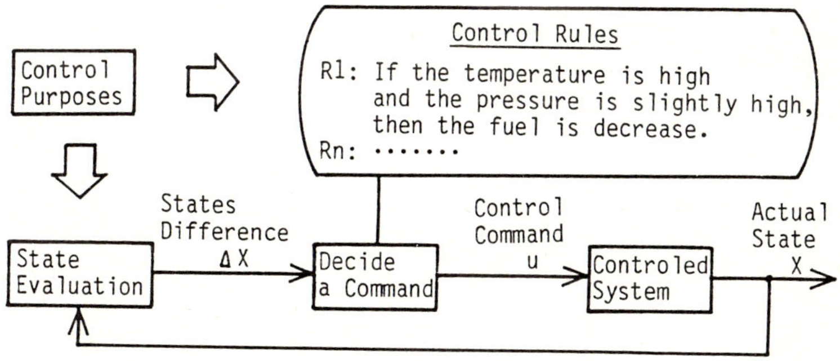

Fig.2 States evaluate Fuzzy control sequence

2.2.2 Predictive (objects evaluation) fuzzy control

A skilled human operator has extensive experience through many experiments with the system's operation. And he can perform high-quality control satisfying the system objectives. However, the above-mentioned fuzzy control cannot evaluate the system objectives.

In order to overcome this problem the predictive fuzzy controller which decides a control action u^* from the objects evaluation of the control results by the control actions has been proposed. In the proposed method,

(i) the control rules **R**(R1,R2···Rn) is described as "Ri: If(u is Ci → x is Ai and y is Bi) then u is Ci",

(ii) the control rule Cj is selected from the predictive results(x,y) that show the highest likelihood.

"When the control notch is not changed, if the train stops in the predetermined allowance zone, then the control notch is not changed." is a typical example of train operation rules. Fig.3 illustrates graphically the control sequence.

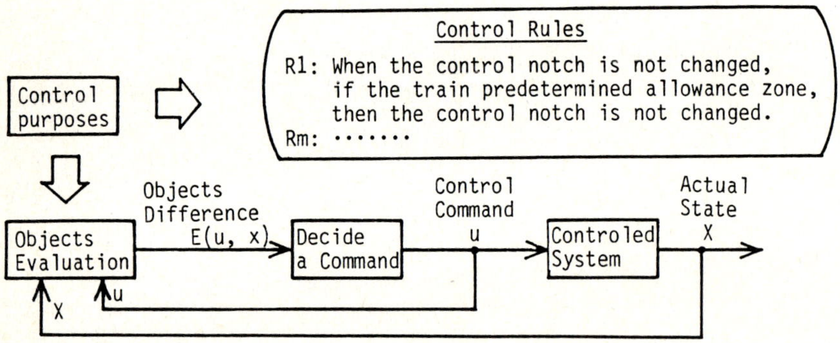

Fig.3 Predictive (objects evaluation) Fuzzy control sequence

The algorithm of the predictive fuzzy controller is as follows:

A fuzzy set **A** on a universe of discourse U is characterized by a membership function μ_A: U ∈(0,1) which associates each element x of U with a number $\mu_A(x)$ in the interval(0,1). A fuzzy set **A** is considered as the union of its constituent singletons. On this basis, **A** may be represented in the form of

$$\mathbf{A} = \int_U \mu_A(x)/x$$

where the integral sign means the union of the fuzzy singletons $\mu_S(x)/x$.

In the predictive fuzzy controller, a value of control command is limited to a discrete number u (u=c1,c2,..,cn), and x and y are assumed to be performance indices for control. Evaluations of x and y, for example "good" or "bad", are defined by fuzzy sets which are characterized by membership functions $\mu Ai(x)$, $\mu Bi(y)$. A fuzzy controller periodically evaluates the efficiency of linguistic control rules such as " If the performance index x is Ai and index y is Bi, when a control command u is decided to be Ci at this time, then this control rule is selected and the control command Ci is decided for output of the controller ". The above linguistic control rule is formulated as follows:

Ri:"If(u is Ci → x is Ai and y is Bi) then u is Ci "

The relationship between x and y of Ri's if part is defined by a membership function of $\mu Pi(Ci:x,y)$ which is shown as a pyramid of Fig.4.

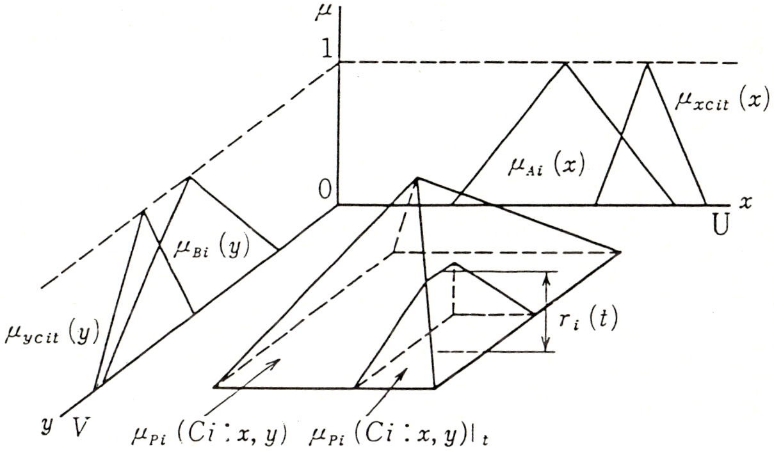

Fig.4 Evaluation of a Fuzzy control rule

When u is Ci at this time t, the fuzzy sets of performance indices X(Ci,t), Y(Ci,t) are predicted by a sub-model of the system. Then, the validity of control rule based on the skilled operator's experiments is given by

$$Pi|t = (Ai \cap X(Ci,t)) \times (Bi \cap Y(Ci,t))$$

and, its value is

$$ri = \sup_{x,y \in U \times V} \mu pi(Ci:x,y)|t.$$

In this way, each control rule is evaluated and the best control rule is selected.

The process to produce the fuzzy logic controller is as follows.

(step-1) Describe human operator's strategies of the system operation

(step-2) Define the meaning of linguistic performance indices

(step-3) Define models to predict the results of an operation

(step-4) Convert the linguistic human operator's strategies into the predictive fuzzy control rules

3. Automatic Train Operation Control And Modelling

ATO is a sub-system of an automatic train control system(ATC). ATC has two subsystems, namely, ATO and the automatic train protection system(ATP) as shown in Fig.5.

The input data of ATO are distance pulses of a tacho-generator, cab signals of the ATP on-board system, position marker detected signals which show that the train has passed a certain position, and supervisory commands from the automatic train supervision system (ATS). The output data of ATO are powering and braking commands to a traction controller and a brake controller.

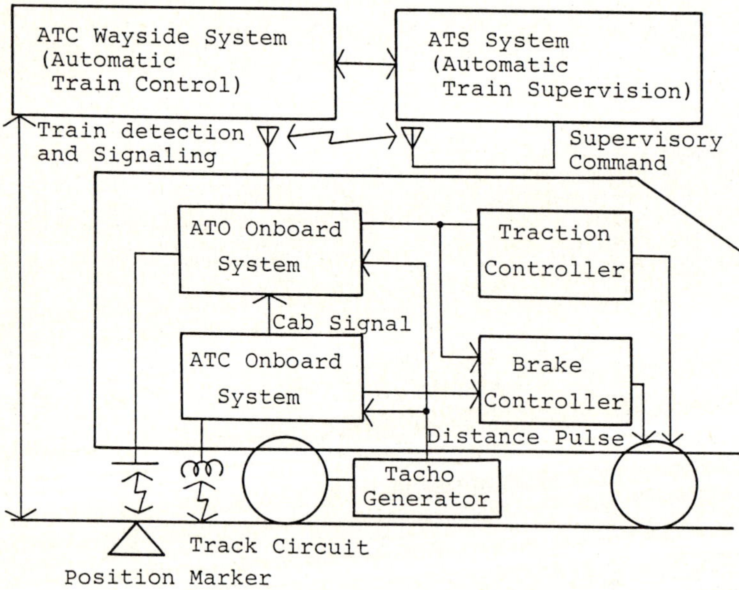

Fig.5 Typical configuration of ATO

A model of the ATO control system is shown in Fig.6. The ATO controller is formulated as follows.

$$\begin{bmatrix} PN(t) \\ BN(t) \end{bmatrix} = A(t,v(t),s(x,t),SPk(t),DT)$$

,where t is time, v is the velocity of train, PN is a power notch (acceleration), BN is a brake notch (deceleration), S is ATP signal, SPk is a point signal of k, and DT is a departure time.

The traction controller and the brake controller are controlled by PN and BN, and these generate the actual power by using those notches.

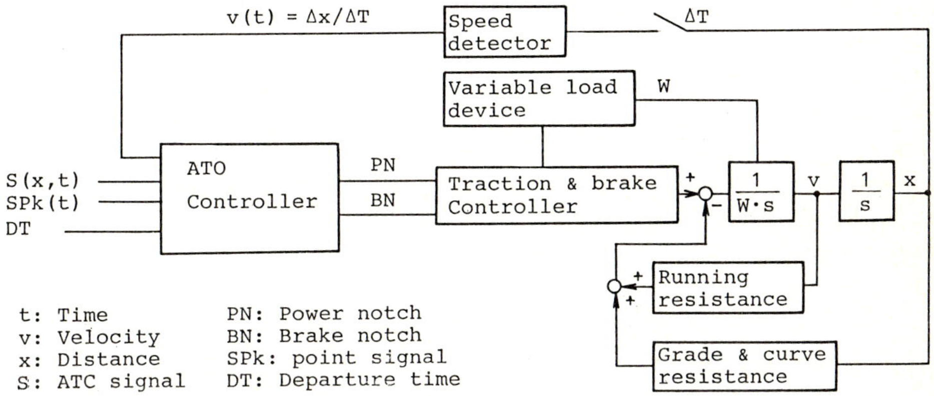

Fig.6 Block diagram of the ATO control system

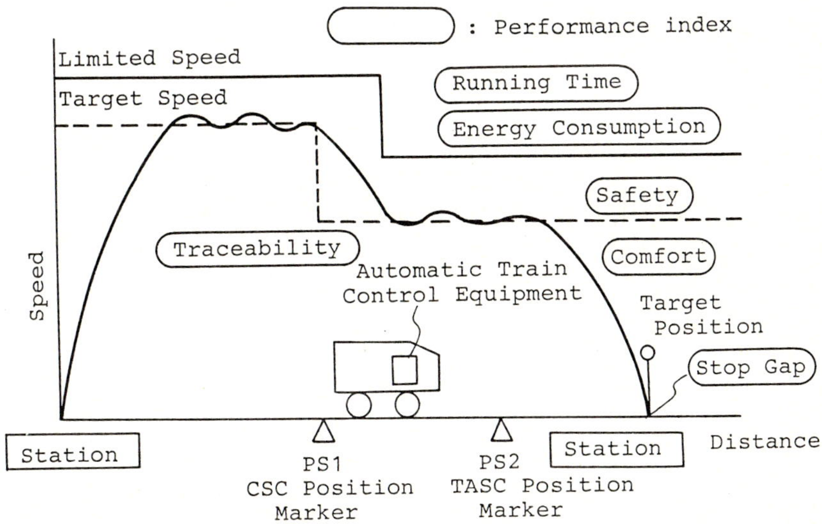

Fig.7 Outline of automatic train operation

4. Human Operator's strategies of Train Operation

The purpose of this paper is to propose an ATO system which realizes train operation similar to a human's operations. As shown in Fig.7, human experts operate trains through evaluating various performance indices such as safety, riding comfort, stop accuracy and so on. The human operator's strategies are divided into two functions. One is the Constant Speed Control (CSC): it starts a train and keeps train speed below the specified one. The other is the Train Automatic Stop Control (TASC) : it regulates a train speed in order to stop the train at a target position of a station. These two functions and a choice of the total control notch are described in the below.

4.1 CSC operation

With the departure command, the operator generates the target speed within the maximum speed limit. If there is a restricted speed track section in front of the train, the target speed is generated at the beginning of this section. In the CSC operation, the control commands are selected by applying experimental control rules in the following way:

(C-1)For safety; if the speed of the train exceeds the speed limit, the maximum brake notch is selected

(C-2)For saving energy; if with coasting the schedule can be kept, coasting is continued.

(C-3)For shortening running time; if the speed is far below the limit, the power notch is selected.

(C-4)For riding comfort; if the train speed is in the predetermined allowance range, the control notch is not changed.

(C-5)For traceability; if the notch is kept constant and the train is about to go out of the allowance range, a $\pm$ n notch is selected so that the target speed way can be traced accurately. In this case, for riding comfort, notch change frequency is considered.

4.2 TASC Operation

An operator starts the TASC operation by detecting the TASC position markers which indicate the distance to the target position of the station. In this control, the control command notch is selected by the following experimental control rules.

(T-1) For riding comfort; when the train is in the TASC zone, and the control notch is not changed, if the train stop predetermined allowance zone, then the control notch is not changed.

(T-2) For minimizing running time and maximizing riding comfort; when the train approaches the TASC zone, the notch is changed from acceleration to a little deceleration by degrees.

(T-3) For the stop gap accuracy; when the train is in the TASC zone, and will not stop within the predetermined allowance zone, a $\pm$ notch is selected so that the train stops accurately at the target position.

4.3 Choice of The Total Control Notch

The traction or brake controller is controlled by the total control notch which is chosen from either the CSC notch or the TASC notch.

5. ATO by Predictive Fuzzy Control

The purpose is to realize an automatic train operation based upon the human operator's experience. Firstly, the fuzzy sets of performance indices such as safety, riding comfort, traceability, energy consump-tion and running time are defined. Secondly, the control rules using these fuzzy sets are described. The symbols which are used below are shown in Table 1.

Table 1 Symbols

t : time (sec)
$x(t)$: location of train (m)
$v(t)$: velocity of train (km/h)
$N(t)$: control command notch
$X(t)$: target position of next station (m)
V_t : target speed (km/h)
T_t : predicted running time (sec)
X_d : forward location where the maximum speed limit is lower (m)
t_s : time to reach X_d point (sec)
X_k : ending location of coasting (m)
$X_z(v)$: beginning point of TASC zone (m)
$t_z = (X_z(v)-x(t))/v(t)$: time to TASC zone (sec)
t_c : elapsed time from last notch change (sec)
N_c : degree of last changed notch
N_p : control command notch to be selected
$V_p(N_p)$: predicted speed when N_p notch is selected (km/h)
V_e : velocity allowance range (km/h)
$X_p(N_p)$: predicted stop position if N_p notch is selected (m)
X_e : allowance of stop gap (m)

5.1 Fuzzy Sets of Performance Indices

To define the performance indices, the three functions denoted below are used.

$$\mathbf{A}(x, a, b) = \begin{cases} b/(a+b-x) & : x < a \\ 1.0 & : a \leq x \end{cases} \quad (b>0)$$

$$\mathbf{A}(x, a, b) = \begin{cases} 0.0 & : x < a+2b \\ 1.0+(a-x)/2b & : a+2b \leq x < a \\ 1.0 & : a \leq x \end{cases} \quad (b<0)$$

$$\mathbf{B}(x, a, b) = \mathbf{A}(2a-x, a, b)$$

$$\mathbf{C}(x,a_1,b_1,a_2,b_2) = \min.(\mathbf{A}(x,a_1,b_1), \mathbf{B}(x,a_2,b_2))$$

(1) Safety performance indices (S)

The safety is evaluated by the time for the train to reach to the lower limited speed (danger) zone.

(a) Danger (SD) : μ_{SD}

$$\mu_{SD}(t_s) = \mathbf{B}(t_s, 0.0, -T_s)$$

(b) Safe (SS) : μ_{SS}

$$\mu_{SS}(t_s) = \mathbf{A}(t_s, -T_s, -T_s)$$

(2) Comfort performance indices (C)

Riding comfort is evaluated by the degree of the last change and the elapsed time from it.

(a) Good comfort (CG) : μ_{CG}

$$\mu_{CG}(t_c, N_c) = \mathbf{A}(t_c, N_c, -N_c+C_b)$$

(b) Poor comfort (CB) : μ_{CB}

$$\mu_{CB}(t_c, N_c) = \mathbf{B}(t_c, C_b, -N_c+C_b)$$

(3) Energy saving performance indices (E)

Energy saving is evaluated by running time margin.

(a) Energy saved running (ES) : μ_{ES}

$$\mu_{ES} = \mathbf{B}(x(t), X_k, -E_x)$$

(b) Not energy saved (consumed) running (EN) : μ_{EN}

$$\mu_{EN} = \mathbf{A}(\ x(t),\ X_k+E_x,\ -E_x)$$

(4) Traceability performance indices (T)

The traceability is evaluated by the difference between the predicted speed and target speed. (Fig.8)

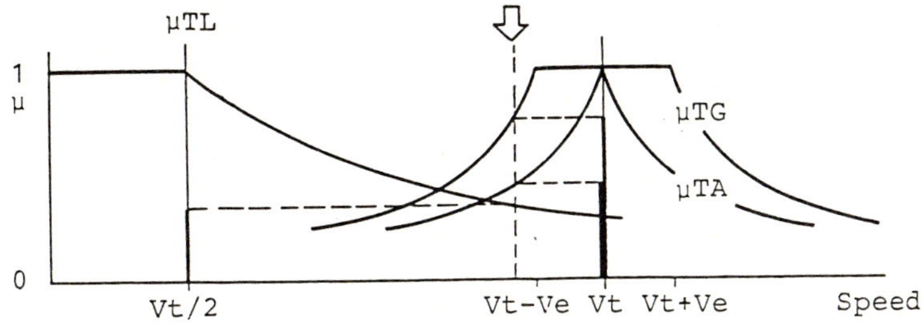

Fig.8 Membership function of the traceability

(a) Good trace (TG) : μ_{TG}

$$\mu_{TG}(V_p(N)) = \mathbf{C}(\ V_p(N),\ V_t-V_e,\ V_e,\ V_t(t)+V_e,\ Ve)$$

(b) Accurate Trace (TA) : μ_{TA}

$$\mu_{TA}(V_p(N)) = \mathbf{C}(\ V_p(N),\ V_t,\ V_e,\ V_t,\ V_e)$$

(c) Low speed (TL) : μ_{TL}

$$\mu_{TL}(V_p(N)) = \mathbf{B}(\ V_p(N),\ V_t/2,\ V_t/4)$$

(5) Running time performance indices (R)

Running time performance is evaluated by a start point for TASC based on the train speed.

(a) In TASC zone (Punctual) (RP) : μ_{RP}

$$\mu_{RP}(t_z) = \mathbf{B}(\ t_z,\ 0.0,\ -R_t)$$

(b) Not in TASC zone (Late) (RF) : μ_{RF}

$$\mu_{RF}(t_z) = \mathbf{A}(\ t_z,\ 2Rt,\ -R_t)$$

(6) Stop gap performance indices (G)

The accuracy of stop gap is evaluated by the difference between the predicted stop position and the target position. (Fig.9)

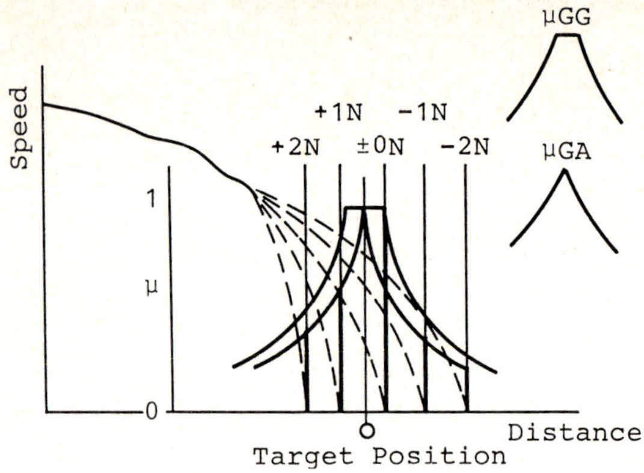

Fig.9 Membership function of the stop accuracy

(a) Good stop (GG) : μ_{GG}

$$\mu_{GG}(X_p(N)) = \mathbf{C}(\ X_p(N),\ X_t - X_e,\ X_e,\ X_t + X_e,\ X_e)$$

(b) Accurate stop (GA) : μ_{GA}

$$\mu_{GA}(X_p(N)) = \mathbf{C}(\ X_p(N),\ X_t,\ X_e,\ X_t,\ X_e)$$

5.2 Predictive Fuzzy Control Rules

The fuzzy control rules have been determined to correspond to the linguistical control rules of human operator's strategies. In the rules, DN is the difference of notches, Pn is the powering notch, and Bn is the braking notch. The maximum members of Pn and Bn are assumed to be 7 and 9.

(1)CSC rules

(C-1)If (N is $(N(t)+B_{max})/2 \rightarrow$ S is SD) Then N is $(N(t)+B_{max})/2$
(C-2)If (N is 0 $\rightarrow$ S is SS, C is CG and E is ES) Then N is 0
(C-3)If (N is P7 $\rightarrow$ S is SS, C is CG and T is TL) Then N is P7
(C-4)If (DN is 0 $\rightarrow$ S is SS and T is TG) Then DN is 0
(C-5)If (DN is n $\rightarrow$ S is SS, C is CG and T is TA) Then DN is n
($n = \pm 1, \pm 2, \pm 3$)

(2) TASC rules

(T-1) If (DN is 0 $\rightarrow$ R is RP and G is GG) Then DN is 0
(T-2·1) If (N is 0 $\rightarrow$ R is RF and C is CG) Then N is 0
(T-2·2) If (N is B1 $\rightarrow$ R is RP and C is CG) Then N is B1
(T-3) If (DN is n $\rightarrow$ R is RP, C is CG and G is GA)Then DN is n
($n = \pm 1, \pm 2, \pm 3$)

6. Realization of the fuzzy ATO controller

The fuzzy ATO control system is incorporated in the simulator for an urban automated guideway transit system (JUMPS; Justified Models for Practical Specification)[12,13]. The JUMPS system calculates the performance of a transit system using graphic display, plotter and printer (Fig.10).

Fig.11 illustrates the structure of the fuzzy ATO algorithm. In the ATO system, the control command is selected by evaluating all of the above-mentioned performance indices rules every 100 ms. An evaluation of the states of the predictive fuzzy controller is displayed on a colour graphic display. Fig.12 is an example of running curves which are drawn by a plotter.

The fuzzy ATO control algorithm (include performance indices and rules) is evaluated by the JUMPS. The algorithm has been built into an on-board controller by a micro-computer(M-6800), as shown in Fig.13.

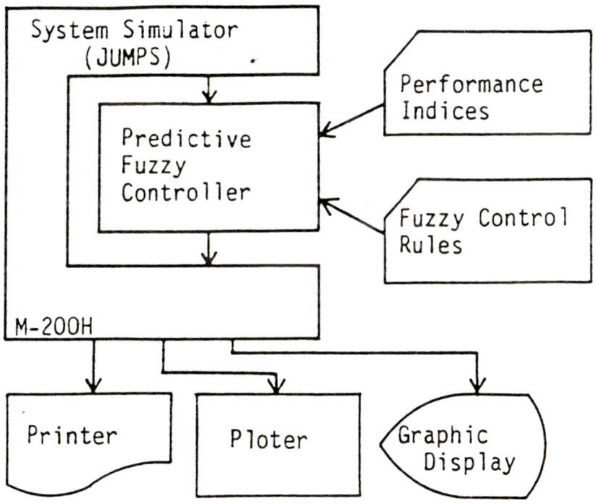

Fig.10 Outline of Fuzzy ATO evaluation system

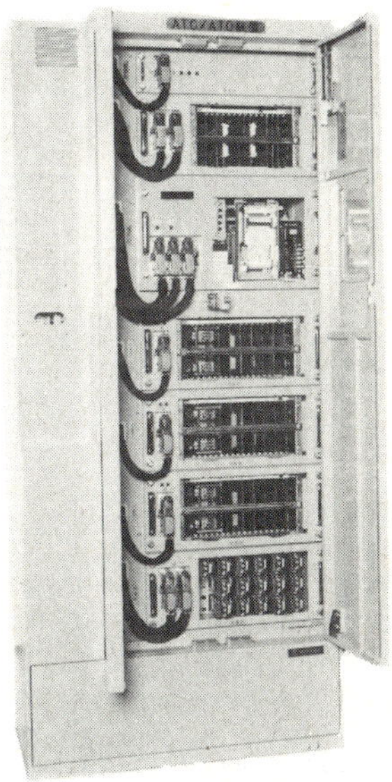

FIG.13 Developed Fuzzy ATO controller

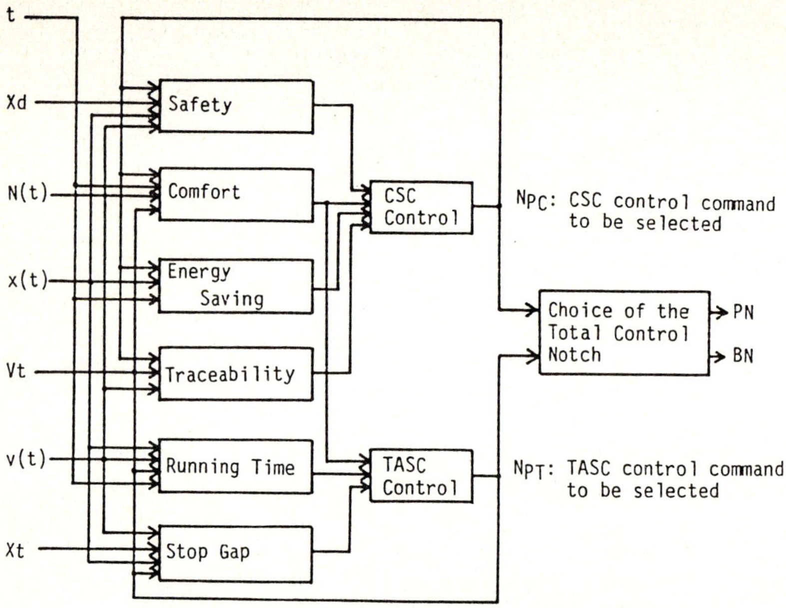

FIg.11 Structure of Fuzzy ATO algorithm

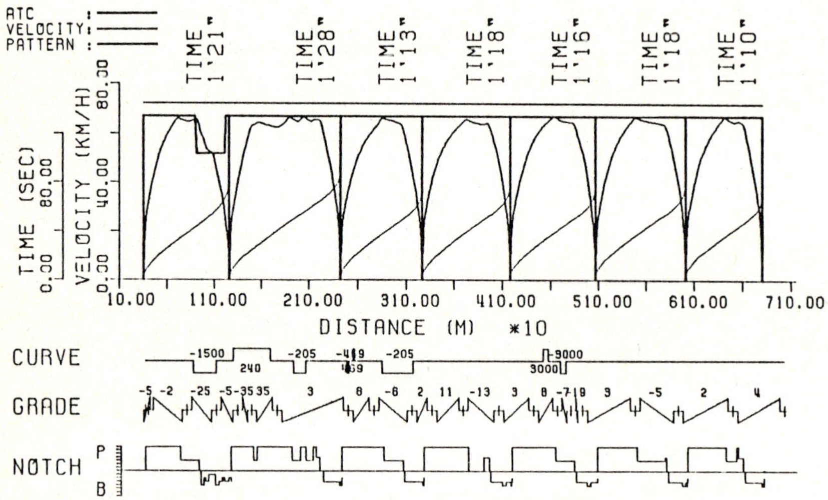

Fig.12 A running curve of the Fuzzy ATO for a subway system by a comuputer simulation

7. Simulations

By the simulation using the JUMPS, the developed predictive fuzzy control ATO (Fuzzy ATO) is compared with the conventional PID control ATO (PID ATO). The parameters used in the simulation are described in Table 2. The simulation results are as follows.

Table 2 Values Used in the Simulation

Train length : 83.5 m
Train weight : 129.0 ton
Running resistance :
$1.97 + 0.016 * v + 0.00084 * v^2$ kg/ton
Number of steps of traction: 7 steps
Maximum acceleration of traction: 3.3 km/h·s
Number of steps of brake: 9 steps
Maximum deceleration of brake B_m:
nominal = 5.14 km/h·s
minimum = 3.6 km/h·s
maximum = 6.68 km/h·s
Dead time of braking system: 0.2 sec
1st order time lag of braking system: 0.6 sec

(a) Riding comfort and stop gap accuracy

Fig.14 shows a summary of simulations for riding comfort and stop gap accuracy. As the results of the simulation, the number of notch changes in the fuzzy ATO is about a half, and the stop gap accuracy is about a third compared

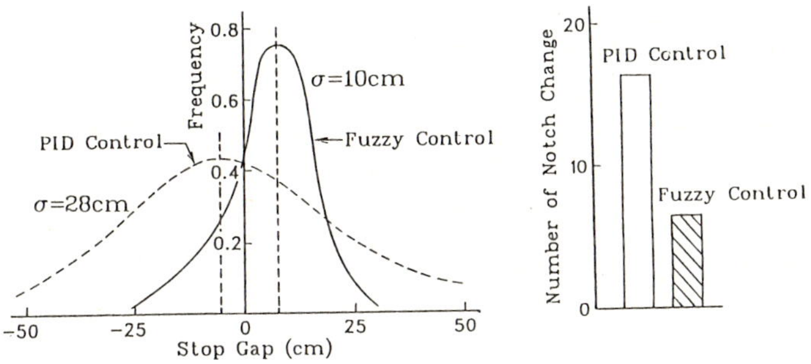

Fig.14 Summary of simulation for riding comfort and stop gap accuracy
- The deceleration to maximum brake notch is 70%, 100% and 130%
- The gradient is $-5\,^{o}/_{oo}$, $0\,^{o}/_{oo}$ (level) and $+5\,^{o}/_{oo}$

with those of the PID ATO. Therefore, the developed fuzzy ATO is able to control a train with good riding comfort and stop gap accuracy.

In the TASC control, the affects of the environment condition, such as track grade and braking ability, are important in controllability. The simulation results, with a grade change just before the station is $0\,^{o}/_{oo}$ to $-20\,^{o}/_{oo}$, are shown in Fig.15. The results obtained are that the stop gap of the Fuzzy ATO is 0.03m and the PID ATO is 1.33m. These results show the higher robustness and control accuracy of the Fuzzy ATO.

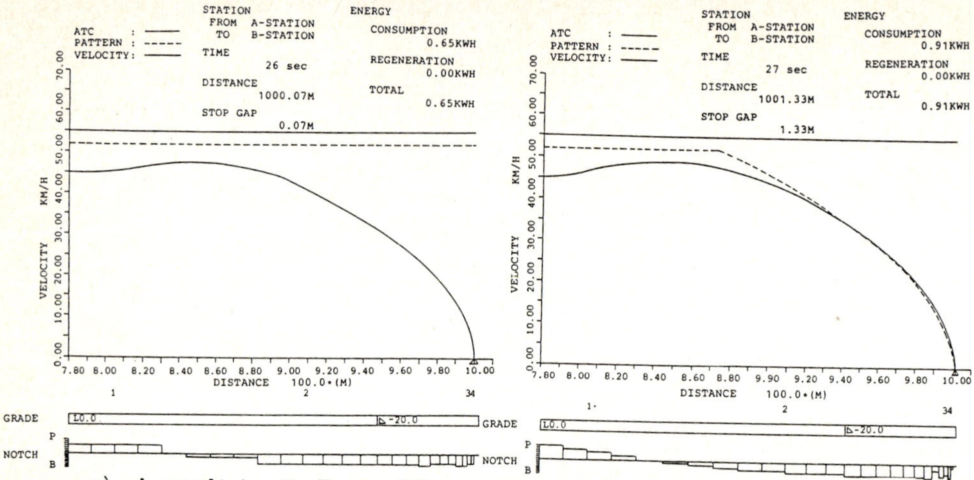

a) A result by the **Fuzzy** ATO b) A result of the PID ATO

Fig.15 Simulation for robust control

(b) Energy consumption

The Fig.16 shows a summary of simulation results on running time and energy consumption. As the results of the simulation, the fuzzy ATO is able to control trains over 10% of energy saving and/or shorten a running time compared with the PID ATO.

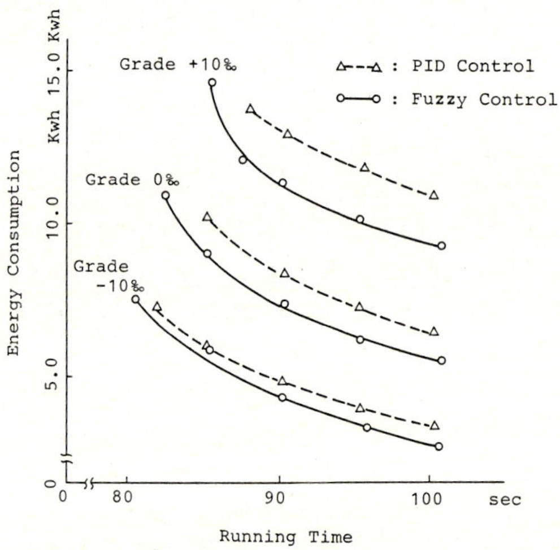

Fig.16 Summary of simulation for energy consumption
- The interval of stations is 1000m
- The gradient is $-10^{0}/_{00}$, $0^{0}/_{00}$ (level) and $+10^{0}/_{00}$

8. Field test for a Subway System

A field test was performed at a subway system. Fig.17 shows a result of the field test. The Fig.12 showed a running curve of a computer simulation for the system, and the results of field test were almost the same as the simulations. The field test shows that the developed Fuzzy ATO controller can operate trains as skillfully as human experts.

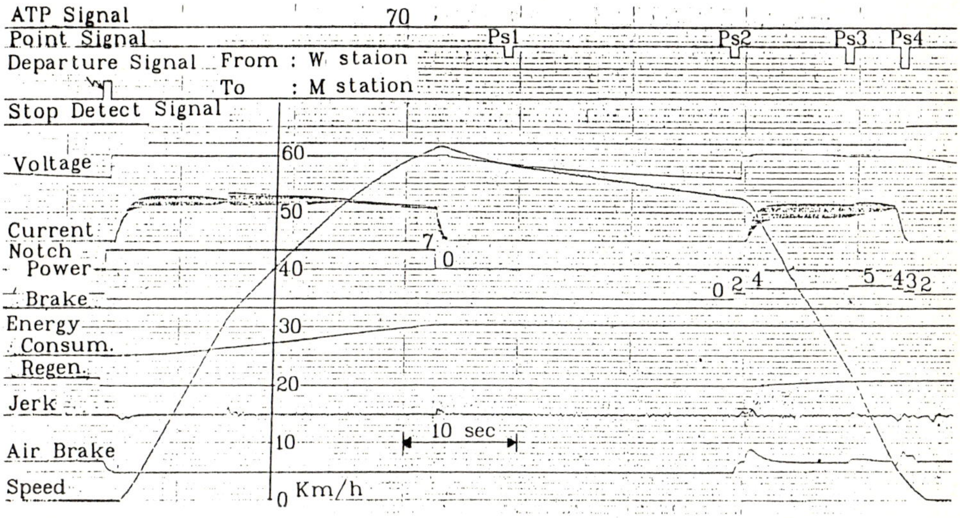

Fig.17 A result of the field tests

9. Conclusions

The predictive fuzzy controller, which selects the most likely control command based on predictions of control results and direct evaluations on the control objectives, has been proposed. This control method has been applied to ATO systems which control trains by evaluating safety, riding comfort, accuracy of stop gap, running time and energy consumption. The simulation studies have revealed that the Fuzzy ATO gives significiant improvements over a conventional PID ATO with respect to the above-mentioned performance indices. Also, it is able to operate robustly under the changing environmental conditions and save over 10% of energy compared with conventional ATO controllers. The proposed Fuzzy ATO algorithm has been built into an on-board controller, and field tests have shown that the developed Fuzzy ATO controller operate trains as skillfully as a human expert. Since control quality is a strong requirement, the proposed predictive fuzzy control is an effective method for overcoming the problems of conventional control systems.

10. Acknowledgements

The authors wish to thank Dr. J. Kawasaki, General Manager and Mr. H. Ihara, Deputy General Manager of Systems Development Laboratory, Hitachi LTD., for continuous support and valuable discussions on this effort. We also wish to thank Mr. T. Takaoka, Senior Engineer of Mito Works, Hitachi LTD., Mr. H. Oshima, Senior Engineer of Industrial Processes Group, Hitachi LTD. and the members of their groups, for their considerable assistance with the development of on-board controller.

References

1. Zadeh,L.A., Outline of a New Approach to the Analysis of Complex Systems and Decision Processes: IEEE Trans. SMC-3-1 (1973) 28-44
2. Mamdani,E.H., Applications of Fuzzy Algorithms for Control of a Simple Dynamic Plant: Proc. of IEEE 121-12 (1974) 1585-1588
3. Pappis,C.P. et al, A Fuzzy Logic Controller for a Traffic Junction: IEEE Trans. SMC-7-10 (1977) 707-712
4. Holmblad,L.P. et al, Control of a Cement Kiln by Fuzzy Logic: M.M.Gupts & E.Sancez (eds.), Fuzzy Information and Decision Process (North-Holland Campany (1982)) 389-399
5. Tong,R.M. et al, Fuzzy Control of the Activated Sludge Wastewater Treament Process: Automatica 16 (1980) 659-701
6. Tong,R.M., A Control Engineering Review of Fuzzy Systems: Automatica 13 (1977) 559-569
7. Yasunobu,S. et al, Fuzzy Control for Automatic Train Operation System: 4th IFAC/IFIP/IFRS Int. Conf. on Transportation Systems (1983) 39-45
8. Miyamoto,S. et al, Predictive Fuzzy Control and Its Application to Automatic Train Operation Systems: The First Int. Conf. on Fuzzy Information Processing July 22-26 (1984)
9. Ishida,S. et al, Train and Traffic Control Systems for Subway: Hitachi Review 25-8 (1976) 283-288
10. Tabata,A. et al, Recent Railway Automation Systems: HItachi Review 27-2 (1978) 85/90
11. Ferbeck,D., The Lille Undergrand - First application of the VAL system: 4th IFAC/IFIP/IFRS Int. Conf. on Transportation Systems (1983) 47-52
12. Miyamoto,S. et al, TRANSPLAN - A Computer Assisted Planning and Designing System for Urban guideway Transit System: SCSC (1978)
13. Miyamoto,S. et al, JUMPS - A Design Support System for Urban Automated Guideway Transit System: ICCS (1978)

Industrial Applications of Fuzzy Control
M. Sugeno (ed.)

APPLICATION OF FUZZY REASONING TO THE WATER PURIFICATION PROCESS

O. Yagishita

Technology Department II, Public Utility Division
Fuji Electric Co., Ltd.
12-1 Yurakucho 1-chome, Chiyoda-ku, Tokyo 100, Japan

O. Itoh

Systems Research Section I, Systems Development Department
Fujifacom Corporation
Fujimachi 1-chome, Hino 191, Japan

M. Sugeno

Department of Systems Science
Tokyo Institute of Technology
4259 Nagatsuta, Midori-ku, Yokohama 227, Japan

In the water purification plant, raw water is promptly purified by injecting chemicals. The amount of chemicals is directly related to the water quality such as turbidity and alkalinity of raw water. At present, however, the process of the chemical reaction to the turbidity has not been clarified as yet, so the amount of chemical cannot be calculated from the data of water quality only. Accordingly, it has to be judged and determined by skilled operators according to the data obtained from past experience. In connection with this, we studied a method to determine the amount of chemicals automatically by reproducing the judgements of the operators by means of fuzzy implication, and performed a field test in the water purification plant using a fuzzy controller. This report deals with the construction an functions of the fuzzy controller and the result of the field test.

INTRODUCTION

The role of a purification plant is to supply high-quality safe drinking water, and water for factories without fail. The source of water for this purpose is underground water and surface water of rivers, lakes, etc. Recently, surface water has been mainly used in order to protect resources. Surface water is available in relatively abundant quantities but contains a large amount of impurities, and that amount fluctuates widely. A coagulant dosage is fed to remove these impurities fully and effectively. The coagulant feed rate has to be changed according to a change in quality of a raw water taken into the purification plant. However, the process of coagulating the impurities with a coagulant is complicated, and only limited items can be measured with on-line water-quality sensors. For these reasons, no practical method of fixing a feed rate has been established yet, and currently, feed rate is fixed by a skilled operator's judgement based on his rich experience.

This paper describes the method of expressing an operator's judgement based on his empirical knowledge by fuzzy reasoning and a coagulant addition fuzzy controller using this method. This paper also reports the result of a field test of this controller at an actual water purification plant.

The feed rate is estimated by the conventional regression model plus fuzzy reasoning, which is used in respect of those areas not handled by the regression model, to estimate the required amount of compensation [2]. A fuzzy implication or the amount of compensation is a combination of the operator's grasp of states (e.g., rise in turbidity and formation of flocs) and the appropriate rate of increase/decrease of the feed rate. Hence the fuzzy controller makes control calculations with the operator's recognition as a input value in addition to process measurements.

The output of the fuzzy controller can be inferred according to compositional rule of inference [1].

R. M. Tong and others incorporated unmeasurable factors of a state into fuzzy reasoning and simulated the active sludge process to demonstrate the effectiveness of the fuzzy controller [3]. The study presented in the paper used a reasoning method based on compositional rule of inference, which was partially improved so as to reduce calculations and thereby make the reasoning readily understandable. This method has already been put into practical use for control of a cement kiln by fuzzy logic [4].

The field test was carried out for two months. Fuzzy implications on the amount of compensation were made on the basis of the operation data for three months including the first one month of the test period. They were verified with the data for the last month of the test period. As a result, it was discovered that the fuzzy controller is effective even in such situation that cannot be handled by the conventional system --- especially a case where the state of flocs must be frequently observed at the time the turbidity of raw water is changing.

Chemicals Addition Control for Purification Plant

The purpose and the method of chemicals addition control at a purification plant will be described here, and the present control method will be explained.

An alkaline agent, coagulant and chlorine agent are fed in the purification plant. Lime or caustic soda is used as an alkaline agent for pH adjustment. Aluminum sulfate or PAC (polymerized aluminium chloride) is used as a coagulant. All of these are used to promote coagulation of the impurities to raise the efficiency of sedimentation.

A suitable chemical agent is selected according to the trend in water quality at the time a purification plant is constructed. Aluminium sulfate is economical but is not very effective in low-temperature water. The effect of PAC is not decreased in low-temperature water, but it is expensive. A chlorine agent is fed for sterilization.

A chlorine agent is called pre-chlorination or post-chlorination agent according to the feed point. A pre-chlorination agent is used to prevent generation of algae or the like in a sedimentation basin, filtration basin, etc., and a post-chlorination agent is used to turn the water into hygienic drinking water.

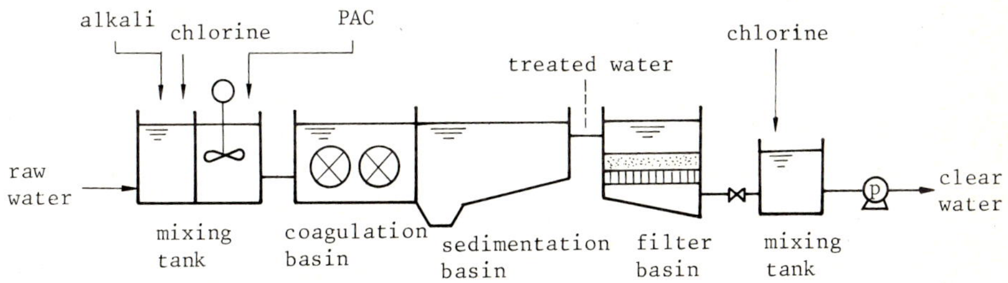

Figure 1
Water purification process

An outline of the purification system and chemicals feed points are shown in Fig. 1. The sedimentation and filtration processes are most fundamental and represent a large portion of the purification plant. The sedimentation basin is designed to precipitate the impurities in water. A coagulant is fed to quicken the sedimentation process. The coagulant aggregates the impurities in water (into flocs) to make them gain weight and sink faster. Thus formation of good flocs is essential for smooth sedimentation, and this gives a reference for coagulant addition. The turbidity of raw water is 10-odd mg/ℓ normally and may rise to even hundreds of mg/ℓ. The impurities are largely removed by a sedimentation basin, and the effluent at its outlet has a turbidity of only several mg/ℓ. The fact that the coagulant plays a very important role can be inferred from this. The residual impurities are then removed by a sand filtration basin to reduce the turbidity further to 1 mg/ℓ or less. If impurities inadequately precipitated are transferred to the filtration basin, that basin is promptly clogged; thus it cannot perform its filtration function.

The method of feed rate control or a chemical agent will be described. A suitable feed rate is prefixed for any particular chemical agent, and the quantity fed is kept in proportion to the flow of the raw water. An alkaline agent and a chlorine agent need to be fed at a constant rate despite some fluctuation of water quality; thus no difficulty is involved in the fixing of a feed rate for such chemicals. On the other hand, a coagulant dosage reacts to the impurities in water through a process so complicated that a process model is hard to work out. For the reason, there is no widely accepted method of fixing a feed rate.

Some of the methods currently used will be explained. One method involves a jar test. Under this method, speciments of raw water are collected in five to six beakers, a coagulant is fed in these beakers at different rates, and the mixtures are stirred. A feed rate is fixed after analysis of the sedimentation of the impurities in these beakers. This is an effective method, but the test takes too much time. The test is especially time-consuming work if the turbidity of the raw water is changing. For the reason, studies have been made to find out a way of fixing a feed rate on the basis of the quality of raw water. Under this method, a regression model for historical plant data is obtained by regression analysis and is used as a feed rate equation. As measurements with on-line water-quality sensors are employed, this method permits automated coagulant addition control. Only limited measurements, however, can be made with on-line water-quality sensors. Since the turbidity is measured by use of transmitted light, particle size and the chemical composition cannot be found out. Thus only gross measurements can be made, and therefore, this method cannot respond to slight changes in water quality.

As can be seen from the forgoing, the main difficulty in coagulant addition control lies in the fixing of an optimum feed rate in response to a change in quality of raw water. an especially important factor is the connection is the coagulant which directly affects the sedimentation process. Hence an analysis will be made in this paper with primary attention to the coagulant.

Application of Fuzzy Reasoning

1. Basic Configuration

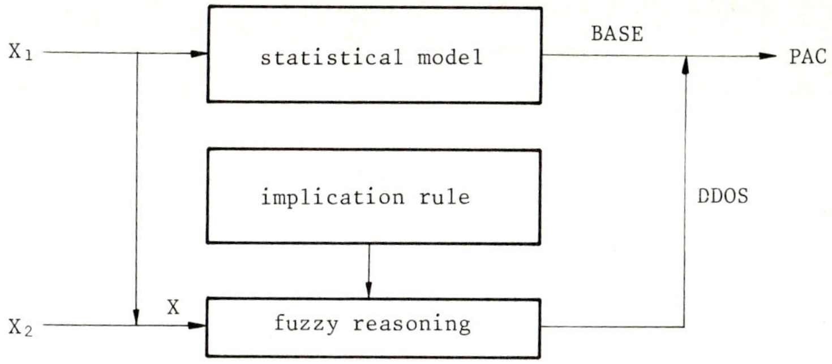

Figure 2
Implication system
X1: water quality data. X2: keyboard input data, PAC: amount of PAC, BASE: amount of PAC basis, DDOS: amount of PAC modification

The basic configuration for estimating the feed rate is shown in Fig. 2. The value of the feed rate equation is used as a reference, and that portion that cannot be handled with that value alone is inferred by fuzzy implication and added as a amount of compensation.

X_1 is the universe of discourse of water quality measurements regularly made and X_2 is the universe of discourse of measurements that cannot be regularly made, such as the state of flocs. The latter measurements are entered by the operator as a "large", "medium" or "small" value. X is the universe of discourse.

A fuzzy implication on the amount of compensation is represented by (1) which is formed of the proposition P that declares the input X to be "large" or "small" and the similar proposition Q for the amount of compensation.

if P then Q (1)
(if P is valid, then Q is valid.)

How P and Q are expressed will be first explained. The proposition P regarding certain input x (εX) is represented by (2) by using the fuzzy variable f which generally declares x to be "large", "small", etc. and the modifier m. F is the universe of discourse of f.

$$P:X = {}^{m}f \quad x \varepsilon X,\ f \varepsilon F \qquad (2)$$

Four cases are considered --- absence of a modifier, excess, ">", falling short, "<", and negation "¬". These are represented by the expressions in (3) below.

$$\begin{aligned} x &= f &&: \text{x is equal to f.} \\ x &= >f &&: \text{x exceeds f.} \qquad (3) \\ x &= <f &&: \text{x is smaller than f.} \\ x &= \neg f &&: \text{x is not f.} \end{aligned}$$

Examples of propositions are given in Fig. 3. As can be seen from Fig. 3, the value exceeding the middle, i.e., the value to the left of the symbol , is of Grade 1. The value to the left of the symbol is of Grade 1. If the fuzzy variable f is "medium" (MM) in ① in Fig. 3 and x is the water temperature (TEMP), the proposition TEMP = ^{m}MM is one among ② to ⑤ in Fig. 3. Each of these is a fuzzy set defined in relation to water temperature.

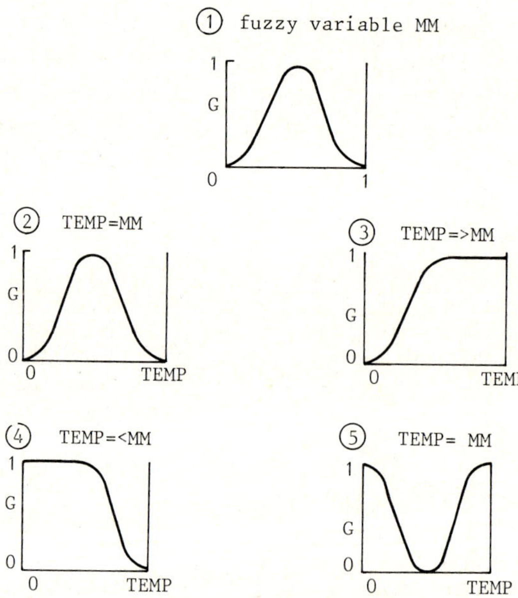

Figure 3

Examples of membership function of modified fuzzy variable

Table 1

Fuzzy variables

variables	support set	μ	σ	meaning
SS	0,1	0.0	0.15	Small Small
SA	0,1	0.0	0.3	Small
SM	0,1	0.3	0.15	Small Midium
MM	0,1	0.5	0.15	Midium
ML	0,1	0.7	0.15	Midium Large
LA	0,1	1.0	0.3	Large
LL	0,1	1.0	0.15	Large Large
NB	-1,1	-1.0	0.4	Negative Big
NM	-1,1	-0.5	0.2	Negative Midium
NS	-1,1	-0.2	0.2	Negative Small
ZE	-1,1	0.0	0.2	Zero
PS	-1,1	0.2	0.2	Positive Small
PM	-1,1	0.5	0.2	Positive Midium
PB	-1,1	1.0	0.4	Positive Big

The fuzzy variable f is represented by a fuzzy set in a unit interval. The membership function hf (x) is represented by (4), and the fuzzy variables shown in Table 1 are provided. "Large", "small", etc. are relative magnitude descriptions. The scope indicated by such a description changes if the scope of the variable changes. Conversely, these descriptions can be generalized by normalizing the range of scope. The variables so obtained are called fuzzy variables. Various inputs can be accommodated only if a number of basic variables are provided.

$$h_f\ (x) = \exp \{- \frac{(x-\mu_f)^2}{2\sigma_f{}^2}\}\ x,\ \mu_f \varepsilon I \qquad (4)$$

The proposition Q regarding the amount of compensation, DDOS, is represented by the expression (5) after the style of handling for (1).

$$Q\ ;\ DDOS = f \quad f\varepsilon F \qquad (5)$$

The fuzzy variable used in this expression is one with the interval $[-1,\ 1]$ on Table 1.

The fuzzy implications on the amount of compensation are given by (6) by use of P represented by (1) and Q represented by (5). The number of inputs is n, and the number of implications m.

$$\begin{array}{lcl} \text{if } P_{11}, \cdots\cdots\cdots, P_{1n} & \text{then} & Q_1 \\ \vdots \quad \ddots\ Pij\ \ddots \quad \vdots & & \vdots \\ \text{if } P_{m1}, \cdots\cdots\cdots, P_{mn} & \text{then} & Q_m \end{array} \qquad (6)$$

2. Fuzzy Reasoning

The amount of compensation is inferred by the fuzzy implications on the amount of compensation given by (6). The propositions P are in AND relation with one another, and the implications are in OR relation with one another. The fuzzy set $\widetilde{DDOS}$ (amount of compensation) is calculated by using (7). This expression simplifies the composition rule of inference. The amount of calculations is small, and the relation between the grade pij of implication and the output pij*Q corresponding to it can be easily grasped. pij represents the grade of the proposition pij, U union of fuzzy subsets, and * multiplication by a constant of the grades of proposition Q. Q_0 indicates that the amount of compensation is "approximately zero". That is, DDOS = ZE. The amount of compensation is zero in a state undefined by the fuzzy implication. This means that the value of the feed rate equation is to be used.

$$\widetilde{DDOS} = U\{(\underset{1\le i\le m}{\text{Min}}\ \underset{1\le j\le n}{pij})\ *Qi,\ p0*Q_0\}$$

$$p0 = 1 - \underset{1\le i\le m}{\text{Max}}\ \{\underset{1\le j\le n}{\text{Min}}\ pij\}$$

Examples are given in Fig. 4. Control rules expressed by fuzzy implications are indicated on the left side.
Their meanings are explained below:

(1) If the rise of the turbidity of the raw water is medium, set the amount of compensation at some positive level.

(2) If the turbidity of the raw water is medium and the water temperature is not low, set the amount of compensation at a medium negative level.

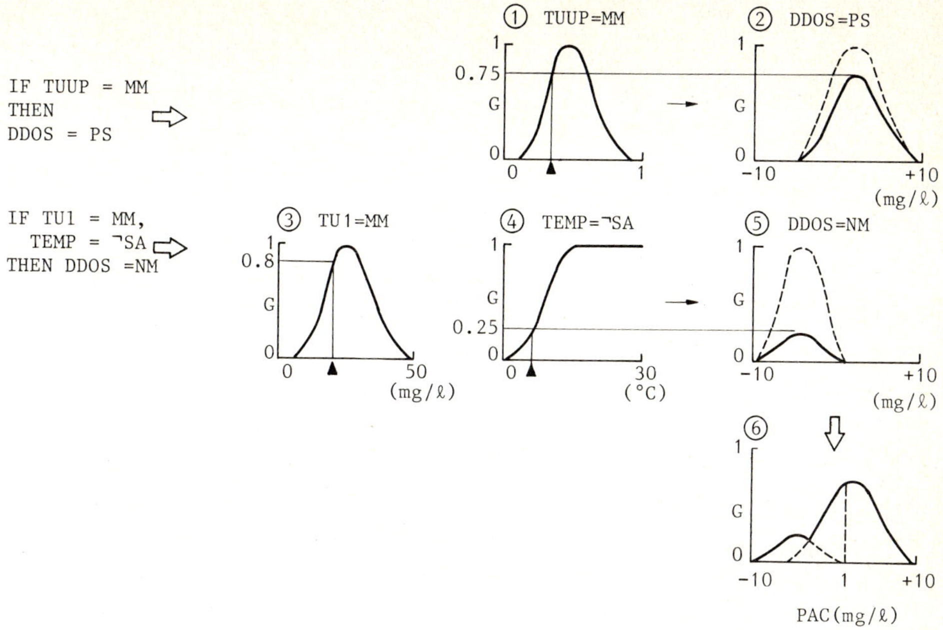

Figure 4
Example of implication

If the measurement and the keyboard entry are at ▲ in the figure, the grade of the proposition TUUP = MM holding true (grade) is 0.75 under the rule (1), and the solid line ② in Fig. 4 is obtained. Under the rule (2), the grades of the propositions are calculated, and the solid line ⑤ in Fig. 4 is obtained. If there are two or more propositions the lowest grade is selected. The functions obtained are combined. The largest function is represented by the solid line ⑥ in Fig. 4. It is a membership function representing a fuzzy set of feed rate compensations. It is abbreviated to h_D. The amount of compensation is derived from h_D by using the expression (8). "d" is a variable within the amount-of-compensation range. That is, the value which divides the area under the membership function represented by ⑥ in Fig. 4 is defined as the amount of compensation. This is because consideration of only the rule having the highest grade is inadequate. The amounts of compensation of the rules have to be reflected according to the grade.

$$DDOS = \int d \, \frac{\Sigma d.hD}{\Sigma h_D} \qquad (8)$$

In the example given in Fig. 4, an extra amount of 1 mg/ℓ is provided.

Composition of Fuzzy Controller

The features of a coagulant addition fuzzy controller include its ability to recognize a process measurement and the operator's recognition of a state such as "large" or "small" flocs, as input and make a control calculation. Such re cognition of a state is based on the operator's long experience, and mere quantification of such recognition may cause loss of an important factor. Hence such recognition by the operator has to be treated as a approximate quantity ("large", "small", etc.), i.e., a fuzzy variable. Moreover, the empirical rule should preferably be related directly to definitions of fuzzy variables, selection of operation instruction variables (turbidity, alkalinity, etc.) and the composition of the implications. It is desirable to make an addition, or change or effect deletion according to the situation so as to facilitate renovation of a fuzzy reasoning. The hardware taken into account is shown in Fig. 5, and the software so designed in Fig. 6.

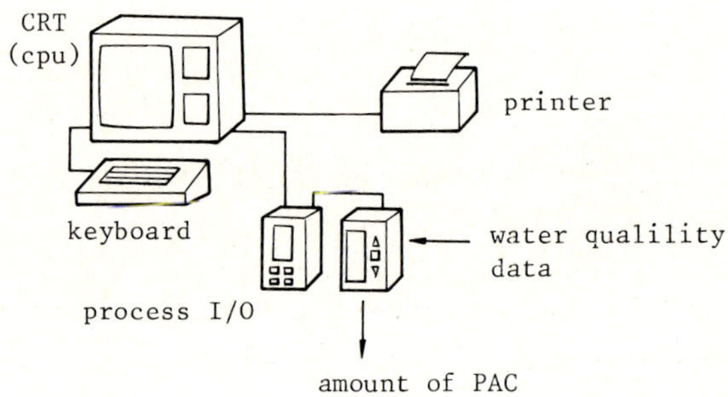

Figure 5
Hardware system

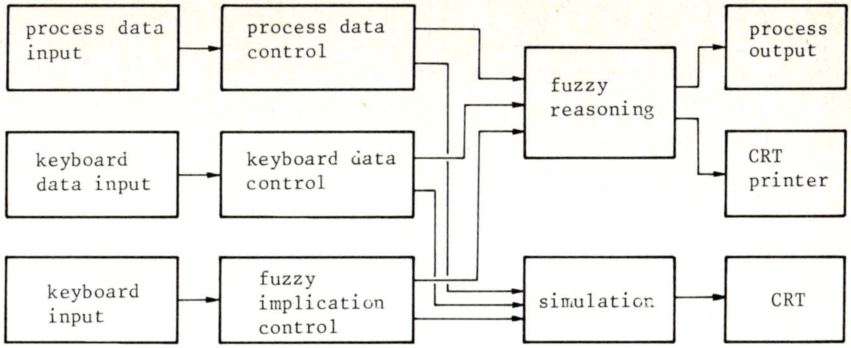

Figure 6
Software system

The functions of the components are described below:

CPU : Calculation of fuzzy logic, simulation

CRT : Displaying of a feed rate, the state of reasoning by control rule, the states of input variables and the result of simulation.

Keyboard : Input of the operator's recognition of a state, fuzzy variables and a change of the implication.

Process I/O : Input of process measurements and output of the reasoning, i.e., the feed rate.

Printer : Output of input data, the state of reasoning by control rule and output of reasoning.

Among these components, the CRT plays an especially important role as a man-to-machine interface. Examples of presentations on the display are given in Fig. 7.

A presentation on the display of a state in shown in ① in Fig. 7. The estimate and the action under the major control rule are displayed. The control rule of the highest grade, that is, the effective rule at that point, is selected and displayed. The grades of the individual proposition of the rule and the grade of that rule are graphically presented in such a way as to permit ready matching with the fuzzy set of the amounts of compensation for reasonings.

The result of a simulation to find out the estimate after a change of the implication rule is shown in ② of Fig. 7. The operator looks at this and decides whether to change the rule, according to the feed rate after that change. The rule can be changed by answering the question "FIX?" with YES.

The section 3 of Fig. 7 is an area where keyboard entries, that is, the operator's recognitions of states as "small", "large", etc., are displayed. In this example, the rise of temperature is small, the flocs are somewhat larger (good) that the average, and the start-up speciality is large.

An addition to and an alternation of fuzzy variables are made in ④ of Fig. 7. General fuzzy variables are listed in Table 1. An addition and an alternation can be made to adapt them to the peculiarities of a place and an individual's peculiar criterion of judgment. A fuzzy set is graphically presented in accordance with given parameters on the left side so that the meanings of those parameters may be readily understood.

In addition to the presentations shown here, there are a total of eleven varieties including presentations of changes in input variables and the feed rate equation and lists of implications, input variables and fuzzy variables.

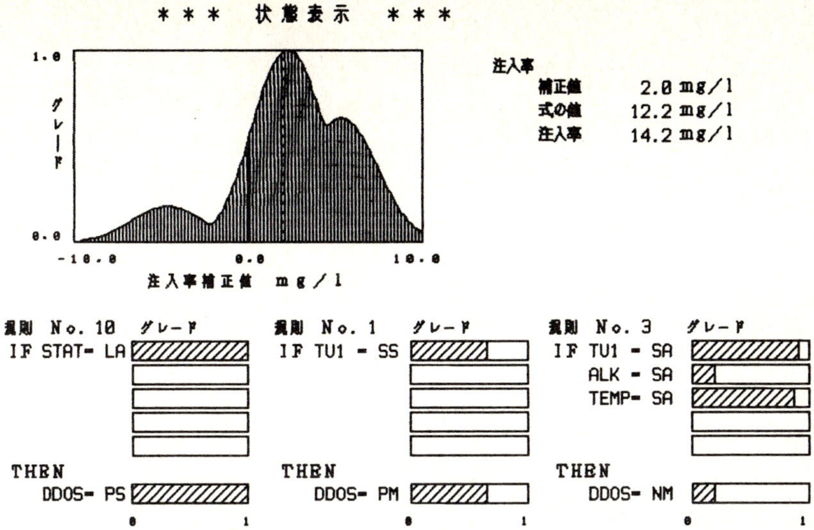

① Result of reasoning

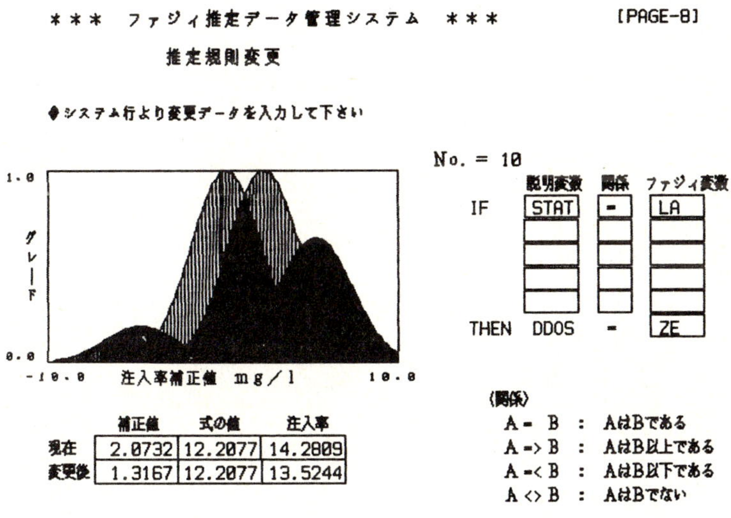

② Simulation and modification of reasoning

＊＊＊ ＫＢ入力データ ＊＊＊

説明変数

No.	変数名	値	コメント
▶ 9	TUUP	▶SA	ﾀﾞｸﾄﾞ ｼﾞｮｳｼｮｳ
10	FLOC	ML	ﾌﾛｯｸ ｼﾞｮｳﾀｲ
11	STAT	LA	ｳﾝﾃﾝ ｶｲｼ
12			
13			
14			
15			
16			

次のファジィ変数を入力して下さい

小さい（少ない） → SA
少し小さい → SM
普通 → MM
少し大きい → ML
大きい（多い） → LA

③ Keyboard input data

＊＊＊ ファジィ推定データ管理システム ＊＊＊ [PAGE-4]

ファジィ変数変更

●システム行より変更データを入力して下さい

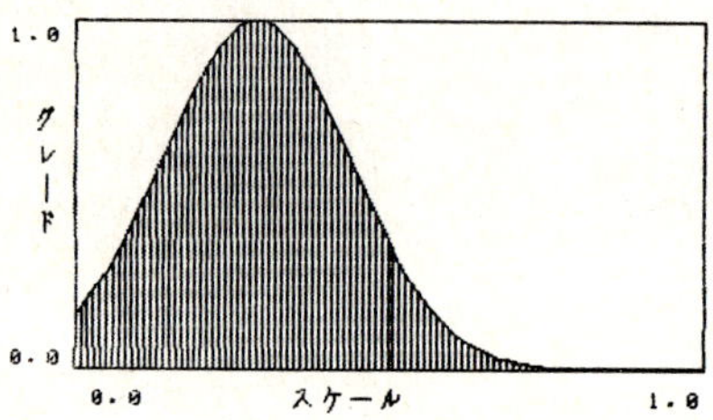

No.	3
▶ファジィ変数名	▶ SM
スケール	0.0 - 1.0
センター	0.3
シグマ	0.15
コメント	SMALL MIDIUM

④ Fuzzy variable modification

Figure 7
Example of CRT output

Field Test

A field test was a made at the Toyoiwa Water Purification Plant in Akita City in October and November in 1983. The test procedure, method and results will be described. PAC was employed at this purification plant as a coaglulant. The plant was in intermittent operation in the daytime only because of secanty raw water supply.

First, the test procedure will be explained. The procedure can be divided into three major parts:

(1) Personal inquiry into empirical rules and drafting of implication.
(2) Varification of implication with actual data and its expansion.
(3) Field test

(1) The results of the personal inquiry will be summarized here. In the directions given below, "addition" means the setting of the feed rate at a level higher by 3 to 4 mg/ℓ, and "subtraction" the setting of the feed rate at a level lower by 3 to 4 mg/ℓ.

. Addition if the turbidity of raw water is rather low.
. Subtraction if the alkalinity is low.
. Addition if the supernatant liquid of the sedimentation basin has a high turbidity.
. Addition if the turbidity of raw water rises (or such rise is anticipated).
. Addition if the flocs are small.
. Slight addition at the time the operation (inter-mittent) is started.

In Table 3, initial implications are made out of the results of the personal inquiry by using output variable symbols in Table 2 and fuzzy variables in Table 1.

Table 2
Variables

	Variables	Symbol	Span	Dimension
-	turbidity of raw water	TU1	0,50	mg/ℓ
	alkalinity	ALK	8,18	"
input data	water temperature	TEMP	0,30	°C
	turbidity of treated water	TUSE	0,3	mg/ℓ
	turbidity increase	TUUP	0,1	-
	floc formation	FLOC	0,1	-
	operation start	STAT	0,1	-
Output data	amount of PAC modification	DDOS	-10,10	mg/ℓ

(2) The field test in October was made under the implications in Table 3. The implications were then expanded by use of the operation data for this period and the data for July and August. At that time, a study was made of various factors including the turbidity of treatment water and the water temperature which were generally believed to affect the feed rate although no such effect had been noted through the operator's personal (firsthand) inquiry. The final implications after the study are shown in Table 4. Fuzzy implications for a case where the raw water has medium turbidity have been added to Table 3. The fuzzy implications for a case where the alkalinity is low were divided into two fuzzy subspaces, and those for a case where the turbidity rises were divided into three fuzzy subspaces.

A mathematical model for the feed rate was worked out by statistical analysis using the data for July, August and October. That equation is given below.

$$\text{BASE} = 18.572 + 2.375\sqrt{\text{TU1}} - 0.93\ \text{ALK} \qquad (9)$$

Where BASE is the reference feed rate (mg/ℓ), TU1 the turbidity of the raw water (mg/ℓ), and ALK the alkalinity (mg/ℓ).

(3) The test for November was carried out with the implications (3) and (2) and the feed rate equation stored in the controller. Details of the installation of the controller are given in Fig. 8.

Next the test method will be described. The test was made by comparison with the value preset by the operator. At a point in the test period, the operator preset feed rate by a jar test or by using his judgment based on experience. It was decided that the validity of fuzzy reasoning be established by demonstrating conformance of the output of fuzzy reasoning to the presetting by a skilled operator and improvement beyond the result obtained with the regression model. The fuzzy reasoning included information regarding the state of flocs and other states unfit for automatic measurement and is expected to provide a better result than the regression mode. By demonstrating that the output of fuzzy reasoning is closer to the result obtained by a skilled operator than to the result obtained with a regression model, it can be proven that the information on non-automatic measurements which is used for fuzzy reasoning plays an important role in fixing feed rate.

Moreover, it can be demonstrated that the experience and knowledge of a skilled operator can be fully utilized, which is the objective of the fuzzy reasoning in this paper, and that by so utilizing a skilled operator's experience and knowledge, a system which can be run by even an unskilled operator with the cleverness of a skilled operator can be realized. Non-automatic measurement information is entered from a keyboard, but during the test, input data were entered when the preset feed rate was changed. This was due to certain judgment of the situation which had led to change of the presetting, and the above set-up was designed to reflect that judgment. There were three operators, and it was discovered as a result of the inquiry that a slight difference in judgment existed among these operators. However, the test period was relatively short, and consequently, such difference for operator to operator could not be varified on the basis of the test results.

Now the test results will be analyzed. As already mentioned, the fuzzy implications established on the basis of the data for July, August and October were applied to the data for November. The effect was varified by comparing the estimate with the number of data collected in November was 122. The value of the feed rate equation along was obtained for comparison. Part of the test results are shown in Fig. 9. It will be seen that the fuzzy reasoning effected close follow-up even in that area which could not be matched with the operator's presetting under the method using a regression model. This can also be demonstrated with the sum of squares of the residuals in Table 5 or the standard deviation of the residual. The residual here refers to the difference between the actual feed rate and the value obtained by fuzzy reasoning and the difference between the actual feed rate and the value of the feed rate equation.

Table 3

Implication rule (primary)

1	IF TU1 =SS	THEN DDOS=PM
2	IF ALK =SA	THEN DDOS=NM
3	IF TUSE=LA	THEN DDOS=PM
4	IF TUUP=LA	THEN DDOS=PM
5	IF FLOC=SA	THEN DDOS=PM
6	IF STAT=LA	THEN DDOS=PS

Table 4

Implication rule (final)

No.				
1	IF TU1 =SS			THEN DDOS=PM
2	IF TU1 =MM	TUSE=¬LA	TEMP=¬SA	THEN DDOS=NM
3	IF TU1 =SA	ALK =SA	TEMP=SA	THEN DDOS=NM
4	IF TU1 =LA	ALK =SA		THEN DDOS=NM
5	IF TUSE=LA			THEN DDOS=NM
6	IF TUUP=LL			THEN DDOS=PB
7	IF TUUP=ML			THEN DDOS=PM
8	IF TUUP=MM			THEN DDOS=PS
9	IF FLOC=SA			THEN DDOS=PM
10	IF STAT=LA			THEN DDOS=PS

Figure 8

Outside view of fuzzy controller installed in the control center of the water purification plant

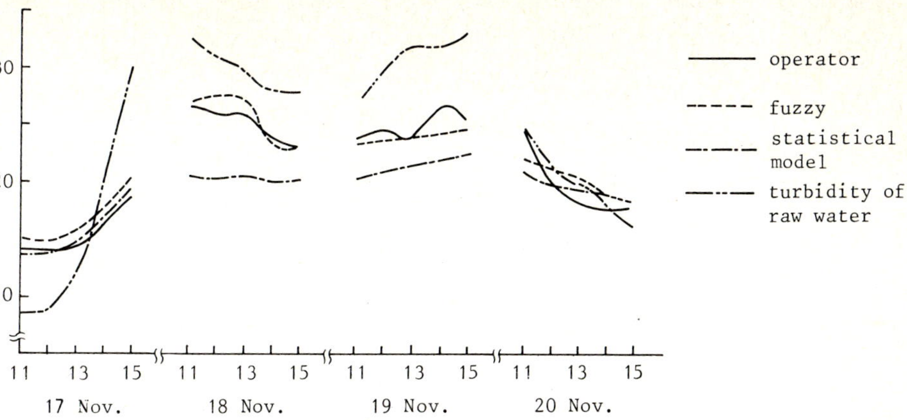

Figure 9
Example of result of fieldtest

Table 5
Example of result of fieldtest

	statistical model	Fuzzy
residual square sum	628	224
standard deviation of residual	2.28	1.36
number of plus Y	33	41
number of minus Y	41	33

Furthermore, the criterion for judgment given by the expression (10) should be considered in order to analyze coagulant addition and its effect. The residual is the value defined already. ΔTUSE is the deviation of the turbidity of the effluent of the sedimentation basin (TUSE) from the average turbidity. To allow for the detention time in the sedimentation basin, ΔTUSE data after time lag of two hours should be used instead of the feed rate.

$$Y = \text{Residual} \times \Delta\text{TUSE} \qquad (10)$$

If Y is positive,

(1) the residual and ΔTUSE are both positive, or
(2) the residual and ΔTUSE are both negative.

If ΔTUSE is positive, the feed rate has to be higher than the actual feed rate. If the residual is positive, this value should be a lowered one. If ΔTUSE is negative, the feed rate should be lowered. In either case, the residual should be a narrowed one. Conversely, if Y is negative, the residual is widened. That is, the greater the number of positive symbols is, the better the feed rate is. As can be seen from this, fuzzy reasoning is superior in this respect also to the method using a regressions model only.

Conclusion

This paper deals with the application of a fuzzy theory to coagulant addition control for a water purification plant. The paper analyzes feed control of a coagulant which is used in larger quantity than any other chemical, and reacts in a complicated process and for which a model cannot be easily worked out. The fixing of a feed rate for the coagulant depends heavily on the operator's judgment based on his experience and knowledge. This judgement is expressed by using fuzzy reasoning so that a feed rate may be fixed by even an unskilled operator. The feed rate was estimated by use of a regression model under the conventional method. The compensation required was calculated and actually made. The necessary amount of compensation was found out by fuzzy reasoning by use of the operator's recognition of a state.

A coagulant addition fuzzy controller incorporating this fuzzy reasoning method was fabricated, and a field test was made at the water purification plant. As a result of this test, is was demonstrated that the output of the fuzzy reasoning is matched with a skilled operator's presetting and is better than the result obtained with a regression model.

A task to be accomplished in the future is addition of a learning system by which the value of feed rate of a chemical agent can be evaluated and the fuzzy implication can be improved as required. Also generalization of rules and development of a general-purpose controller will be in increasing demand in future.

Before concluding the paper, we wish to express our sincere gratitude to the staff of the Toyoiwa Water Purification Plant in Akita City in Japan for their kind cooperation in the field test.

References

1) Michio Sugano: "Application of Fuzzy Set and Logic To Control", "Measurement and Control", Vol. 18, No. 2, PP. 150 to 160 (1975)

2) Yagishita and: Tanaka "Estimation of Chemicals Feed Rate By Fuzzy Reasoning", 34th National Water Service Research Presentation and Lecture Meeting, PP. 365 to 367 (1983)

3) R. M. Tong et: al Fuzzy Control of the Activated Sludge Water Treatment Process; Automatica, Vol. 16, PP. 695 to 701 (1980)

4) L. P. Holmblod and: J. J. Ostergaard Control of a Cement Kiln by Fuzzy Logic, in Fuzzy Information and Decision Processors (M. M. Gupta and E. Sanchez eds.), North-Holland, PP. 389 to 399 (1982)

Industrial Applications of Fuzzy Control
M. Sugeno (ed.)

THE APPLICATION OF A FUZZY CONTROLLER TO THE CONTROL OF A MULTI-DEGREE-OF-FREEDOM ROBOT ARM

E.M. Scharf & N.J. Mandic

Department of Electrical & Electronic Engineering
Queen Mary College
University of London

This paper describes the application of a self-organising fuzzy controller to one, and then simultaneously, to two movements of a commercially produced revolute joint robot arm. Results are presented of step response and tracking tests for both cases; these results are often superior to those obtained using a conventional PID controller.

Die Anwendung eines selbst-organisierenden Fuzzy-reglers an ein, und dann gleichzeitig, an zwei Antriebe eines kommerziell erzeugten Roboters wird hier beschrieben. Ergebnisse der Sprungcharakteristik- und Gleichlaufexperimente werden für beide Fälle präsentiert; diese Resultate erwiesen dass der Fuzzy-regler oft einem konventionellen PID-regler überlegen war.

INTRODUCTION

Robots with revolute joints are widely used throughout industry since they are fast acting, simple to construct and approach the flexibility of the human arm. Manufacturing output can be raised and the economics of each robot can be improved by increasing the robot's operating speed. However, the moment of inertia about each of the main control axes of the robot will change both with the relative disposition of its links and with the load carried at its tip. At high speeds this results in very complex dynamics [3, 18] and implies correspondingly complex control schemes.

A model reference scheme for robot control has been demonstrated with the aid of a robot arm simulation by Dubowsky [2]; a second order equation is used to describe the dynamics of each joint; the algorithms for each joint do not interact computationally with each other. According to Leininger [7] this method does not work for manipulators with gravity dependent terms; in his paper, Leininger describes his adaptation of Wellstead's pole placement self-tuning algorithm to a simulation of the JPL-Stanford arm. Vähä and Halme [17] describe an adaptive multi-variable controller which operates in conjunction with a stochastic linear state model of the dynamics of the manipulator joints; this model is based on the manipulator kinematics and its parameters and states are estimated on-line by a non-linear filter. Hewitt [4] has looked at the decoupled dynamic control of robot arms and Richards [12] is at present investigating multivariable adaptive control methods.

In this paper we consider the practical application of a much simpler heuristic algorithm to the simultaneous control of the shoulder and arm movements of a commercially produced robot arm; this algorithm is based on linguistic rules and an associated inference mechanism. These rules may be ported to a computer environment by the use of Fuzzy Logic whose principles were first enunciated by L.A. Zadeh [20]. Professor Mamdani and his co-workers then forged the link between fuzzy-logic and control engineering [8, 1, 11] during the 1970s.

An important industrial application arising from this link occurred in the control of cement kilns, notably in Denmark [16, 9, 6, 5]. The controllers employed for these systems were of the non-learning variety, operating according to a fixed set of linguistic rules which embodied the experience which the kiln operators had gathered over many years. The cement manufacturing process has time constants ranging from tens of minutes to tens of hours; the rule-based controller had a sampling time of several minutes and acted in a supervisory capacity, supplying the necessary setpoints to much faster, but un-intelligent, local PID controllers initimately connected with the kiln hardware such as the drive motors.

The robot controller differs from this former application in four respects. (1) It forms its own rules without even the need for any safety priming rules; thus the controller can adapt to changes in robot configuration and loading. (2) The dynamics of the robot necessitate the much shorter sampling times of 20 to 40 milliseconds; this means that the execution times of the individual software modules can become significant in relation to the sampling period despite the inherent simplicity of the algorithm and the use of 16-bit microprocessors. (3) In order to cope with the changing robot dynamics, each algorithm needs to act in the direct forward path of its respective motor control loop, as opposed to supplying setpoints to the PID motor control loops. (4) The output from a controller is not fed to an accumulator (integrator) but is interpreted directly as the width of the motor drive pulse since the joint dynamics already include the integration effect; this is so since the motor angle is the integral of motor speed which itself approximates to a second order Laplace Transfer Function of the motor drive voltage.

THE NON-LEARNING CONTROLLER

The rule-base or decision table, of which Fig. 1 is an example, contains linguistic rules of the form:

IF (A AND B) THEN C.

In the present context, the antecedents A and B refer to linguistic statements about the error between setpoint and actual robot link angle and the rate of change of this error. Each non-zero entry in the decision table corresponds to one rule; the two axes, or universes of discourse, of the decision table are for change-in-error (CE) and error (E) respectively. Each axis has 13 entries from -6 through 0 to +6. The points 1 to 6 can be taken to mean: very small, small, medium, quite large, large and very large. Thus if error and change-in-error are respectively positive small and negative very small we consider that the rule for E = +1 and CE = -2 is completely fullfilled with a degree of fulfillment D = 1. The fuzzy subset of each antecedent (E or CE) has, for computational simplicity, a triangular shape with D-values of 0.3, 0.7, 1.0, 0.7 and 0.3 although other shapes such as Gaussian could be used. Hence, for the particular rule in question, we could, for example, have:

E	CE	D
+1	-2	1.0
+1	-1	0.7
+3	-2	0.3
+3	-5	0.0 .

The AND connective is thus interpreted geometrically by the three dimensional intersection (union or minimum (MIN)) of the solid projections of the fuzzy subsets of E and CE parallel to the CE and E axes.

The consequent of each rule in the matrix has one of 13 values -6 to +6, represented here, for software reasons, as 1 to 13; this is thus the width of the combined output universe of discourse of all the consequents. Each consequent represents the contribution of the particular rule to the output action set (eg. fig. 2); this in turn provides the width of the drive pulse to the appropriate

E \ CE	-6	-5	-4	-3	-2	-1	0	+1	+2	+3	+4	+5	+6
-6	.	.	.	.	.	.	-6	.	.	.	.	.	.
-5	.	.	.	.	.	-6	.	.	.	.	.	.	.
-4	.	.	.	.	.	.	.	.	.	.	.	.	.
-3	.	.	.	.	.	.	-3	.	.	.	.	.	.
-2	.	.	.	.	.	.	-5	+1	.	.	.	.	.
-1	.	.	.	+1	+1	-1	-2	+2	.	.	.	.	.
0	.	.	+2	+4	+3	.	0	+5	+5	.	.	.	.
+1	.	.	.	.	+5	.	.	+6	.	.	.	.	.
+2	.	.	.	.	.	+6	.	.	.	.	.	.	.
+3	.	.	.	.	.	+6	+6	.	.	.	.	.	.
+4	.	.	.	.	.	.	.	.	.	.	.	.	.
+5	.	.	.	.	.	.	.	.	.	.	.	.	.
+6	.	.	.	.	.	.	.	.	.	.	.	.	.

Fig. 1 A Typical Rule Base (Decision Table)

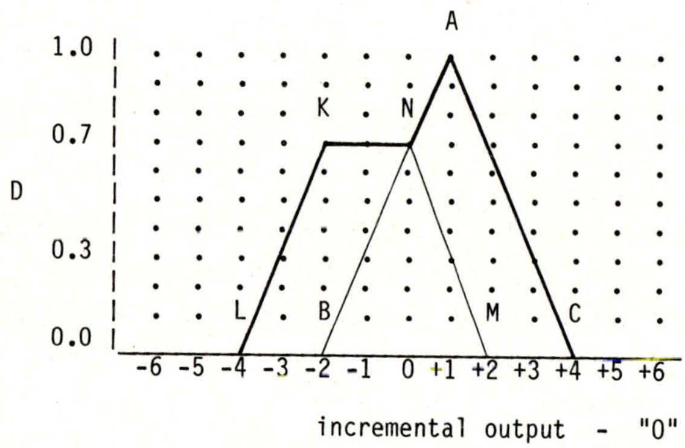

Fig. 2 Superimposing Rules - The ELSE Connective

E \ CE	-6	-5	-4	-3	-2	-1	0	+1	+2	+3	+4	+5	+6
-6	0	0	-1	-2	-3	-4	-6	-6	-6	-6	-6	-6	-6
-5	0	0	0	-1	-2	-3	-4	-4	-5	-5	-6	-6	-6
-4	0	0	0	0	-1	-2	-3	-3	-4	-5	-5	-6	-6
-3	+1	0	0	0	0	-1	-2	-2	-3	-4	-5	-5	-6
-2	+2	+1	0	0	0	0	-1	-1	-2	-3	-4	-5	-6
-1	+3	+2	+1	0	0	0	-1	-1	-1	-2	-3	-4	-5
0	+4	+3	+2	+1	+1	0	0	0	-1	-1	-2	-3	-4
+1	+5	+4	+3	+2	+1	+1	+1	0	0	0	-1	-2	-3
+2	+6	+5	+4	+3	+2	+1	+1	0	0	0	0	-1	-2
+3	+6	+5	+5	+4	+3	+2	+2	+1	0	0	0	0	-1
+4	+6	+6	+5	+5	+4	+3	+3	+2	+1	0	0	0	0
+5	+6	+6	+6	+5	+5	+4	+4	+3	+2	+1	0	0	0
+6	+6	+6	+6	+6	+6	+6	+6	+4	+3	+2	+1	0	0

Fig. 3 The Performance Index Matrix (Yamazaki [19])

motor as described below.

The rule in fig. 1 for E = -1, CE = -2 would contribute the shape ABC to the output action set in fig. 2. The rule for E = -1, CE = -1 would contribute the truncated triangular shape KLMN to fig. 2 since we use a MIN (minimum) [8, 11] operation to represent the THEN connective. We could also scale down the original triangle [19] to give centre heights of 0.7 and 0.3 as appropriate, but this would result in increased processing time.

The collective effect of all the rules can be given as:

(rule 1) ELSE (rule 2) ELSE (rule 3) ELSE ... .

As indicated by fig. 2, the ELSE connectives superimpose the contributions of each rule to form the net output action set. ELSE can be regarded as an INCLUSIVE OR function or (geometrically) as a MAXIMUM (MAX) function. Thus for E = -1, CE = -2, we have the shape ANKLC providing the geometrical representation of:

(rule at E=-1, CE=-2) ELSE (rule at E=-1, CE=-1) .

To obtain a unique output one may find the centre of gravity of the output action set [19]. It has been found by the authors that this does not yield such a fast response as, but tends to give a more smoothly varying output compared with, the mean of maxima procedure [11] which equates the output to the position half way between the two extreme maximum peaks of the action set. The considerations of the previous paragraph together with the implementations of the THEN and ELSE connectives form the substance of the rule inference mechanism.

LEARNING

Learning is done by reference to an incremental performance matrix which has the same size and axes as the rule matrix. This matrix has integer entries ranging from -6 to +6; these entries being the reinforcements, positive or negative which are given to the rule corresponding to a given (E,CE) pair in the rule matrix; reinforcement also implies rule creation. Fig. 3 shows the matrix due to Yamazaki [19] although the version due to Procyk and Mamdani [11] has also been used. The performance matrix is derived from linguistic statements of the form given above.

Learning proceeds as follows [11]. Present E and CE values point to a value P in the performance matrix. U is the incremental controller output which is subsequently converted to a measure in milliseconds of the width of the motor drive pulse.

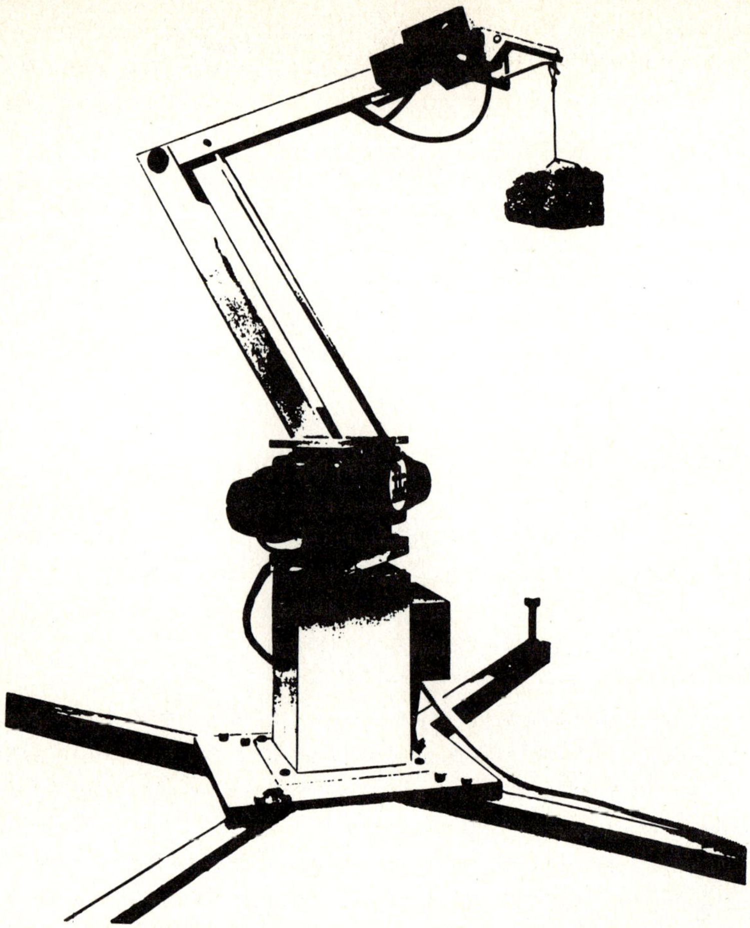

Maximum lifting capability:	1.5 kilogrammes
Length of shoulder link:	0.57 metres
Length of arm:	0.43 metres
Maximum slew speed:	35 deg/sec (approx)
Estimated Dominant Time Constant:	90 milliseconds

Fig. 4 The Robot Arm

Nominally ten past states, each consisting of the triplet (E,CE,U), are stored in a first-in-first-out "history" buffer. Assume (1) a delay of m samples through the process being controlled, (2) that the present E and CE values are mainly due to the output U_m m samples ago and (3) U_m was mainly caused by the rule in the rule matrix at (E_m,CE_m) where (E_m, CE_m, U_m) comprise the state m samples ago. Then the rule entry for (E_m, CE_m) in the rule matrix becomes (U_m + P), whatever the rule entry may have been up to the present time. In order that the modified learning process due to Yamazaki [19] could also be tried, the entry for U_m in the first-in-first-out buffer is also updated.

The self-organising system (SOC) comprises both the more conventional rule-based controller and the above learning process.

EXPERIMENTAL CONSIDERATIONS

The physical details of the robot are summarised in fig. 4 and its main movements - rotation, shoulder, arm - are each interfaced to the DEC LSI 11/23 processor as described below.

The D.C. servo motors of the robot are rated at 1/12th H.P. and, for speed and efficiency, their armatures are each driven by a nominal 60 volt pulse using a 4-transistor bridge. Pulse widths are either in the range 3 to 19.5 or 6 to 39 milliseconds, to fit in with loop sampling times of 20 or 40 milliseconds respectively.

For economy, angular feedback is derived from shaft-mounted servo potentiometers with 0.1% linearity. A 12-bit A/D convertor provides a resolution of 18 bits per degree. Each angle is read 8 times in succession to reduce the effect of noise to +/- 1 bit. Noise filtering is also aided by evaluating the rate of change in error over four samples.

The LSI 11/23 processor runs under the real-time operating system RT-11; the software is coded in a mix of assembler for time-critical portions and the Pascal-like process control language RTL/2.

ALGORITHMIC CONSIDERATIONS

The system proposed for each of the three main movements is outlined in fig. 5.

In the input stage the error between setpoint and robot joint angle is formed and, as previously stated, the rate of change of error is evaluated from error values taken over four consecutive samples so as to provide noise filtering. Both error

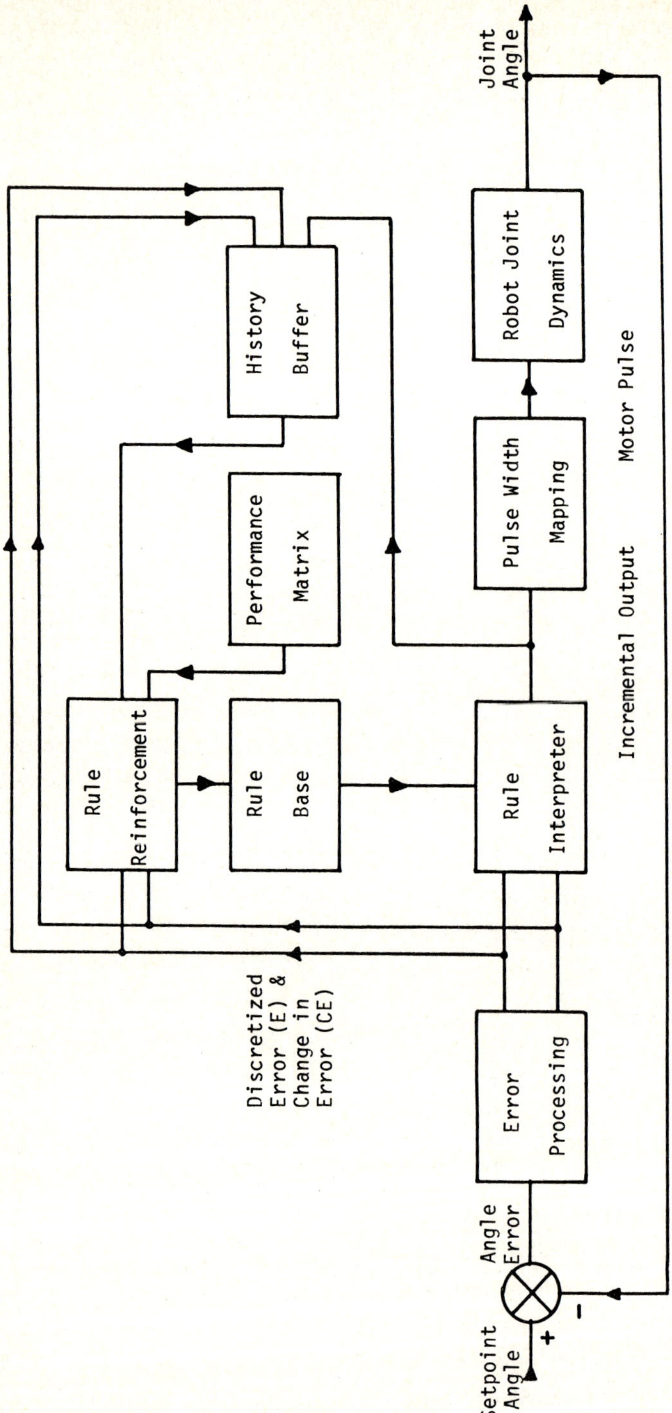

Fig. 5 The Self-Organising Robot Controller

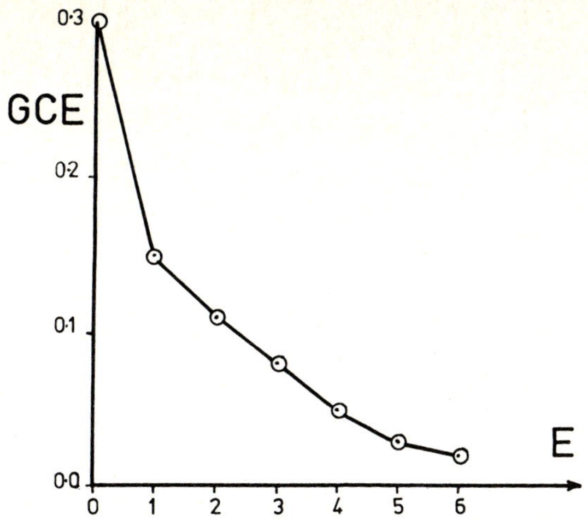

Fig. 6 Functional Dependence of GCE on E

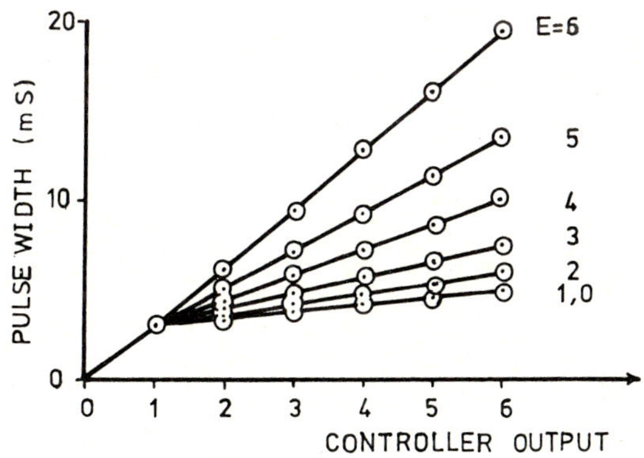

Fig. 7 Motor Pulse Width as a Function of
E and Incremental Controller Output

and change in error are then multiplied by appropriate gains (GE and GCE), limited and quantized, each to one of 13 levels, to give two discrete quantities E and CE, each with a range -6 through 0 to +6. Since the damping of the process response is proportional to the magnitude of GCE [11,8], GCE was made functionally dependent on the discretized error E (fig. 6); this achieved a fast rise time together with good damping close to the set-point. E and CE is then used to index the rule base and hence the output action set may be formed as already described.

A finer controller output action near the setpoint is achieved by using a non-linear mapping, (fig. 7), which is also a function of the discretized error E. This mapping converts the discrete integer controller output values (range -6 to +6) to forward or reverse motor pulses of widths between 3 and 19.5 or 6 and 39 milliseconds as required.

Separate analysis [15] on the shoulder movement showed that this presented to its drive motor a process whose dominant time constant was approximately 90 milliseconds. This suggested that a sampling time of 20 milliseconds would be feasible for the SOC algorithm.

RESULTS

Introduction

Pick-and-place action of the robot was investigated by angular step response experiments; tracking tests were made with welding and machining requirements in mind. Of the three main robot movements, rotation, shoulder and arm, the latter two were considered in order to maximise the possible variation in moment of inertia.

One Degree of Freedom

These results were detailed in an earlier publication [13] by the authors but the load cycling tests should be mentioned here since they involve two movements. The performance matrix due to Procyk and Mamdani [11] was used here.

The shoulder was driven by SOC and the arm was driven seperately (anti-clockwise) by a servo-motor controller chip (type NE544) as shown in fig. 8k. Table 1 shows the experimental loading sequence. The results in figs. 8a-h show the detailed angular response near the setpoint in each case. The first 8 results show the effect of load cycling on the fuzzy controller with c and f being non-learning runs (ie. the previous set of rules was employed). The tip load was 1.035 kg.

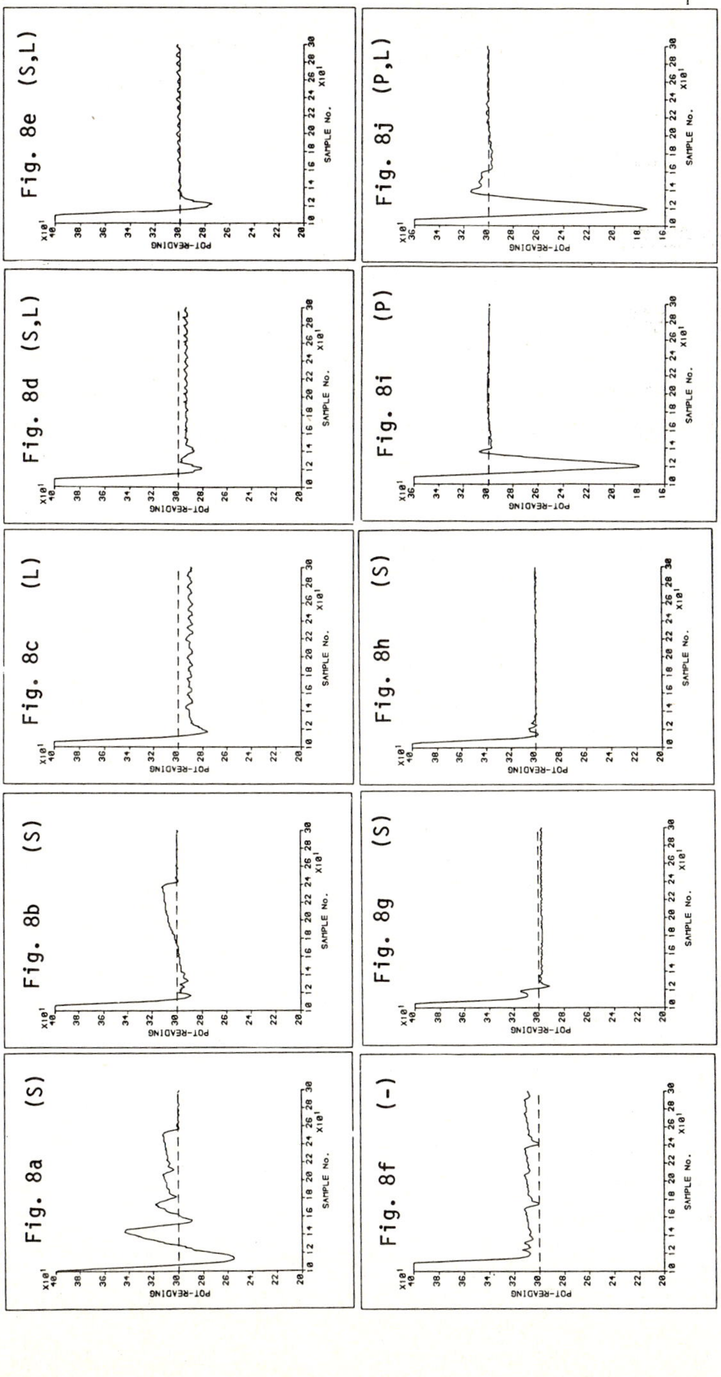

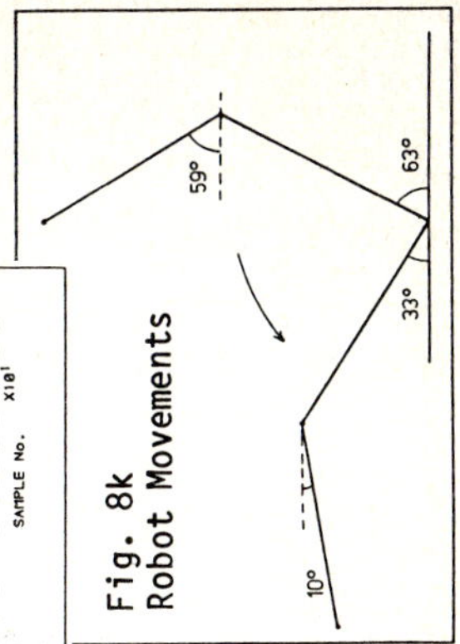

Fig. 8 a-k Load Cycling Experiments (S=Learn (SOC), L=load, P=PID)

Note how the response improved with learning (eg. figs. 8a and 8b) as well as the the offset (figs. 8c and 8f) which appeared when learning was inhibited.

The results for PID control are shown in figs 8i and 8j and show how the dynamics of the robot arm system do change when a load is added. The internal PID gains are 2.0, 0.15, 1.0 respectively and were established with some difficulty to achieve an overshoot of approximately 20% when the robot was carrying no load.

Table 1

Details for Load Cycling Tests

fig	controller	learn?	load
a	fuzzy	y	0
b	"	y	0
c	"	n	y
d	"	y	y
e	"	y	y
f	"	n	0
g	"	y	0
h	"	y	0
i	PID	-	0
j	"	-	y

Experiments with a Two Degree of Freedom System

Introduction

The robot control was decoupled into two separate SOC systems applied respectively to the shoulder and arm movements. Each SOC system thus had to learn its own rules and follow its own setpoint, irrespective of any cross-coupling effects imposed by the dynamics of the robot arm itself. The sampling time of the whole software was kept to 20 milliseconds. This was achieved by writing the computationally intensive routines such as the controller and input modules in assembler as opposed to the high level RTL/2 language.

For this set of experiments, the output action set was processed by the centre-of-

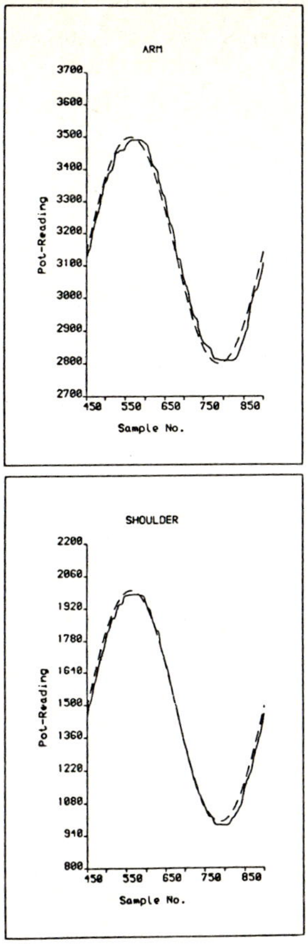

Fig. 9a
SOC (Variable O/P Gain)

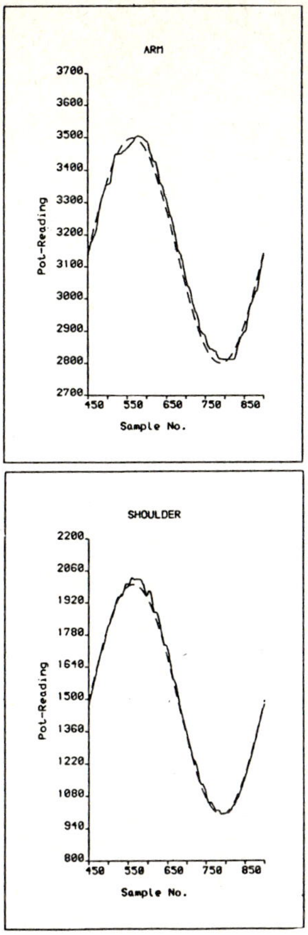

Fig. 9b
SOC (Constant O/P Gain)

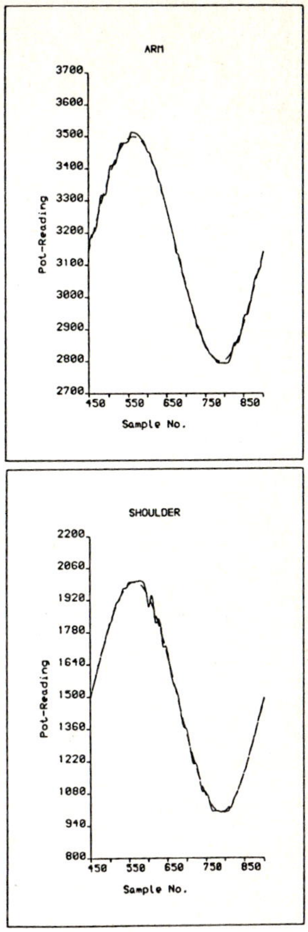

Fig. 9c
PID (Continuous Input)

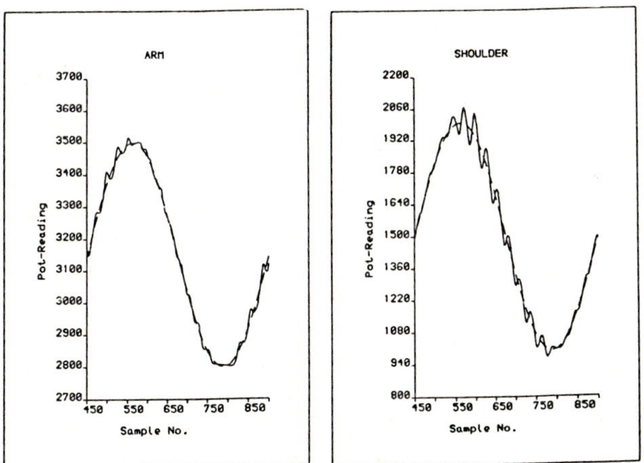

Fig. 9d
PID (Quantised Input)

Fig. 9a-e
Sinusoidal Tracking - I

Fig. 9e
Robot Movements

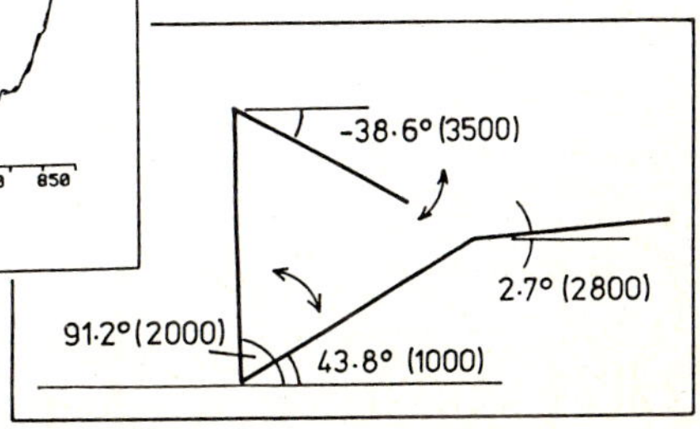

gravity method, and a continuous controller output was used. The performance matrix was due to Yamazaki [19]. Learning took place throughout a single experimental run as opposed to the separate runs in the load cycling experiments. The second set of results for sinusoidal tracking together with the results for following a square trajectory have been reported in an earlier paper [10] but are repeated here for completeness.

Sinusoidal Tracking - 1

These experiments considered the angular responses of shoulder and arm to sinusoidal inputs each with a period of 9 seconds and are summarized in table 2. Each of the figures 9a to 9d show angular responses for both arm and shoulder movements. The responses are given for samples 450 to 900 since by this time some learning will have taken place for the fuzzy controllers. Fig. 9e shows the range of angles through which the shoulder and arm traversed, (potentiometer readings, in brackets, relate the angles to figs. a to d).

Table 2 - Sinusoidal Tracking - 1

fig	description
a	Fuzzy - CG variable medium output gain
b	Fuzzy - CG constant maximum output gain
c	PID - "continuous input"
d	PID - "quantised input"

The responses of fig. 9a are smoother and have a smaller tracking error than the responses of fig. 9b for which the corresponding system has consistently the maximum output gain relating controller output action to motor pulse width (cf. fig. 7). The "continuous input" PID controller has responses comparing favourably with those of the fuzzy controller of fig. 9a; the "discontinous input" PID of fig. 9d shows considerable oscillations about a mean value and compares unfavourably with fig. 9a, even though its input error information is quantized to the same level (+/- 0.5 degree) as with the fuzzy controller of fig. 9a.

Sinusoidal Tracking - 2

In this case shoulder and arm movements were required to track sinusoids with periods of 9 and 4 seconds respectively over a range of angles shown in fig. 10c, (potentiometer readings are in brackets). The higher frequency for the arm was intended to present a significant disturbance to the shoulder. A mean-of-maxima

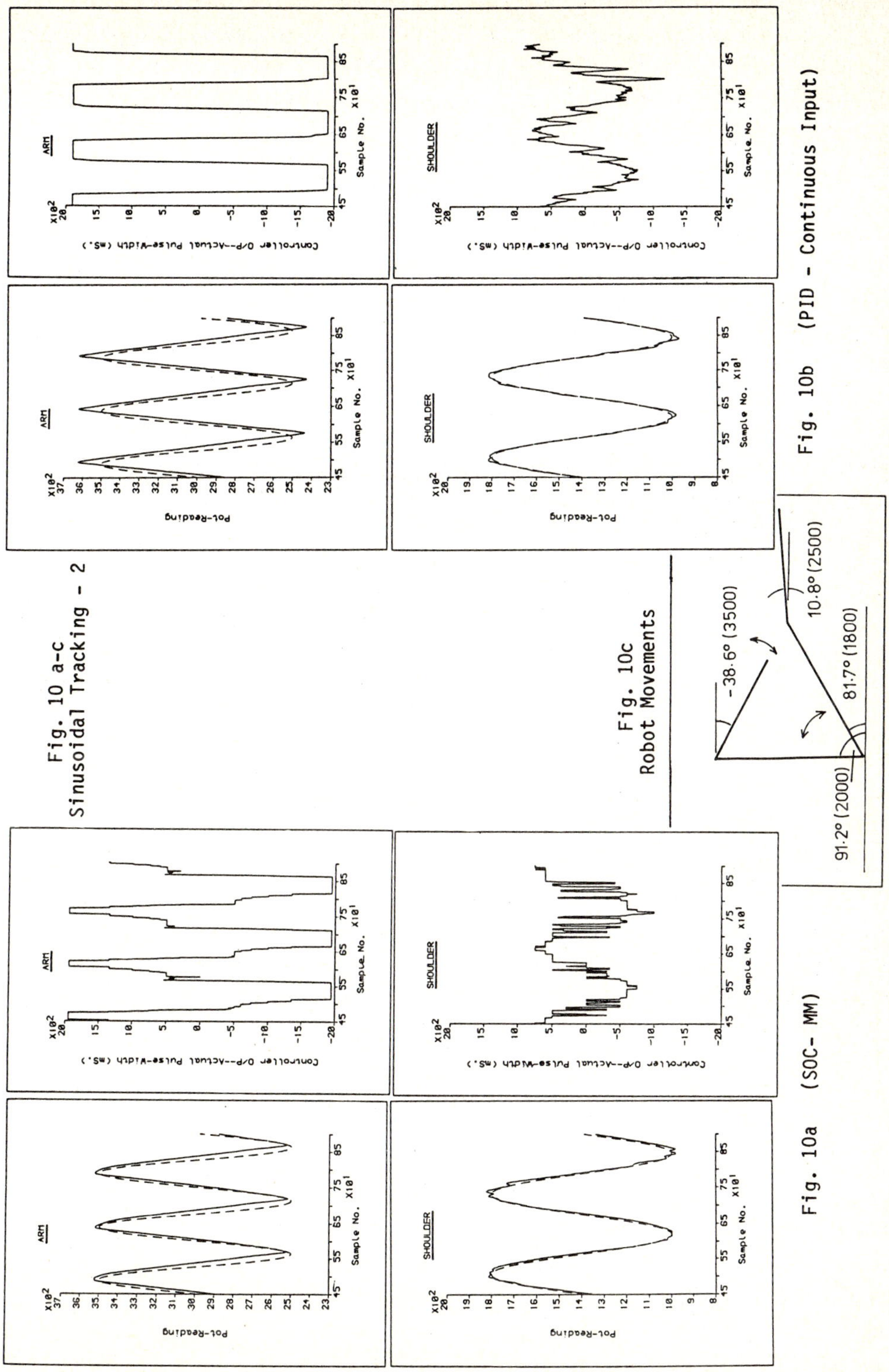

Fig. 10 a-c
Sinusoidal Tracking - 2

Fig. 10a (SOC- MM)

Fig. 10b (PID - Continuous Input)

Fig. 10c
Robot Movements

fuzzy controller was used to generate the results of fig. 10a and a continuous PID controller for fig. 10b. Each of the figures 10a and 10b shows the angular response and controller outputs (in terms of motor pulse widths in milliseconds) for both arm and shoulder motions. Again, responses are shown for samples 450 to 900 since it assumed that by this time the fuzzy controller will have done some learning.

The shoulder responses in both cases show comparable error and smoothness; however, the arm response of the fuzzy controller tracked the 4 second period sinusoid much more closely than the PID (with its triangular response) and the arm controller output did not saturate to the extent that the PID output to the arm did. The fuzzy controller was thus better able to predict the fast changes at the extremes of the input demand sinusoidal waveform.

Tracking a Square

For these results, a kinematic transform, based on the length of the shoulder and arm links, was used to create the demand angle vs. time relation for each of the arm and shoulder motors. These demand curves caused the tip of the arm link to traverse a square of side 40 cm in a z-x plane and centred at a distance x = 55cm and z = 55cm from the origin of this plane; this z-x plane had its origin at the support end of the shoulder link. Each figure (11a,b,c) shows two successive traverses of the square, the starting point being outside the trajectory. Each square was traversed in 9 seconds, and the maximum deviations in x and z directions for the second square of each experiment are summarised in table 3.

The fuzzy controller (fig. 11c) has a similar performance to the continuous input PID controller (fig. 11a) but shows a slight lag on the right-hand down-stroke of the square; the maximum z-deviation for the PID controller is about twice that of the fuzzy controller. With an identical quantisation level of the input error of 1 degree, the fuzzy controller (fig. 11c) has a performance superior to that of the PID controller (fig. 11b).

Table 3 - Square Tracking

Experiment	x-deviation	z-deviation
a: continuous PID	1.97 cm	1.97 cm
b: quantised PID	3.50 cm	2.30 cm
c: fuzzy (CG)	1.97 cm	0.87 cm

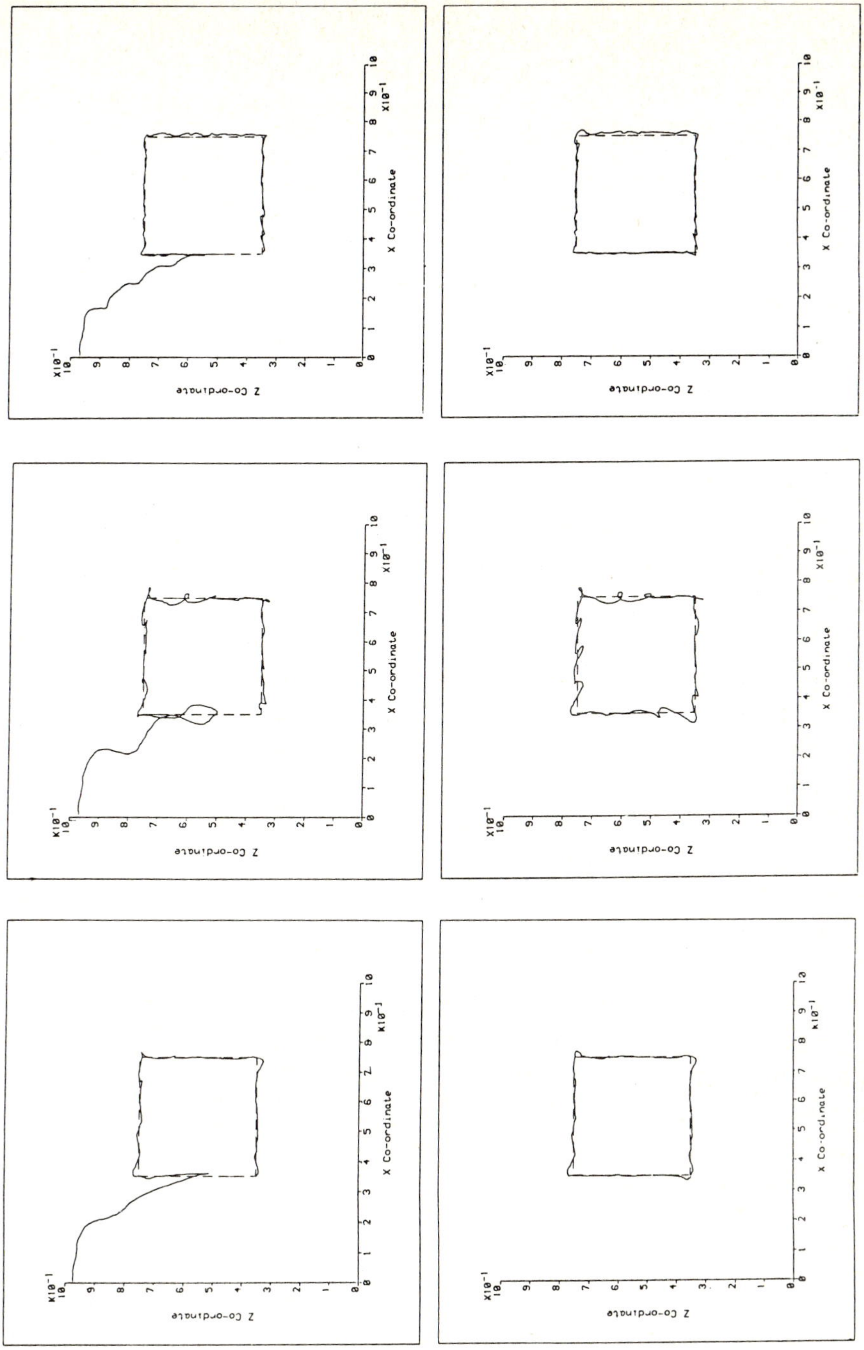

Fig. 11a Continuous PID

Fig. 11b Quantised PID

Fig. 11c SOC (CG)

Fig. 11 a-c Tracking a Square

DISCUSSION

The results, especially for the tracking tests, showed that SOC could give similar or even superior results to those of the continuous input PID controller. This showed that SOC also possesses an in-built filtering effect. The coarseness of the information reaching SOC gives the algorithm, we believe a robustness with respect to noise although it will adversely affect the system performance near the set-point. The ability of SOC to deal with transient effects was demonstated, especially by the load cycling results.

The formation of the rules in the 13x13 rule matrix should also be considered. Rule saturation would occur when a rule had reached one of the limiting values of -6 or +6. The "mean-of-maxima" (MM) method tended to cause a process response which was less smooth than that obtainable with the "centre-of-gravity" (CG) method; this was ascribed to the fact that rule saturation occurred closer to the centre (E=0, CE=0) of the rule matrix and that there was a starker transition between neighbouring rules. Results (eg. [14]) indicated that the MM method yielded a better transient performance, the CG method a better steady state performance.

By expressing the change-in-error (GCE) gain as a non-linear function of the quantised input error (E), rule formation could be kept away from the extreme edges of the rule matrix; this was found to improve the performance of the controller.

The robot arm is a real system and not a just a simulation; this had the advantage of providing a practical application for a fuzzy controller but meant that results were not precisely repeatable. The contructional imperfections of the robot will have influenced the results. The back-lash associated with each of the three movements was at least one degree; also the arm would move by one degree, when in a nominally horizontal position, when the shoulder moved from a vertical position to a position ten degrees off horizontal - this may have influenced the square tracking experiments. It was estimated that the variation that could be achieved in moment of inertia with respect to the shoulder motor was about 10% and some variation is indicated by figs. 8i and 8j. A more enhanced variation would be more representative of most industrial robots and thus provide a better test bed for SOC.

The use of an integration at the controller output as well as the use of a PI combination instead of the more conventional PD combination were both tried but found to give results inferior to those obtainable with the present configuration. In both cases, the system became less stable. In one instance, the sign of the error derivative was inverted - the system eventually achieved a satisfactory performance during step response tests but it took about twice as many learning runs to do so.

Future work will look at improvements to the learning mechanism and at the use of rule interpolation so as not to discard any input error information as at present; it is hoped that the latter measure would allow finer control close to the setpoint. We also wish to investigate the use of the multi-input-multi-output form of SOC. Some tuning of the controller gains GE and GCE as well as the gain and output characteristics was found to be necessary and methods of expediting this process should also be considered. Application to a more demanding industrial robot is also planned.

CONCLUSIONS

This work has demonstrated the feasibilty of applying a self-organising fuzzy algorithm in the direct forward path of a robot movement control loop to deal with the complex high-speed dynamics associated with robot motion. SOC has shown a performance which is comparable to, and often even superior to, a conventional PID controller despite the fact that SOC discretizes the input error information and thus works with less input information than its traditional rival. This became particularly evident in the square and sinusoidal tracking experiments which were performed with welding and machining requirements in mind. Step response investigations were undertaken with pick-and-place requirements in mind and demonstrated the ability of SOC to deal effectively with the transient part of the response, being able to achieve satisfactory performance after usually only three learning runs; SOC exhibited rise times which were usually equivalent to, or slightly in excess of, those obtained with a PID controller.

The mean-of-maxima method yielded a fuzzy controller whose setpoint and transient performance was less smooth but otherwise preferable to that obtained using the centre-of-gravity method. This could be related to the starker variations between neighbouring rules in the rule-matrix, especially near the setpoint, resulting from the former method.

Suggestions aimed at improving the learning and fine control performance have been given; future plans also include the implementation of a multi-input-multi-output form of SOC.

ACKNOWLEDGEMENTS

We are very grateful to the British Science and Engineering Research Council which has supported one of the authors (N. Mandic) and provided the equipment for this work through grant number GR/B 7940.0. We would like to thank Professor E.H. Mamdani for his useful suggestions during the project and we are also indebted to the staff of the mechanical and electronics workshops in our department for the assistance they have rendered at various stages of the project.

REFERENCES

[1] Assilian S. and Mamdani E.H., An Experiment in Linguistic Synthesis with a Fuzzy Logic Controller, Int. J. of Man-Machine Studies, 7 (1974), 1-13.

[2] Dubowsky S., On the Adaptive Control of Robotic Manipulators: The Discrete-Time Case, Proceedings of the Joint Automatic Control Conference, (Charlottesville, Virginia, U.S.A., 1981).

[3] George Lee C.S., Robot Arm Kinematics, Dynamics and Control, IEEE Computer, (December 1982).

[4] Hewitt J.R., 1980, Dynamic Decoupled Control of Robot Arms, Colloquium on Control of Manipulators and Robotic Devices, (IEE, London, England, April 1980).

[5] Holmblad I.P. and Ostergaard J.J., Fuzzy Logic Control: Operator Experience Applied in Automatic Process Control, FLS Review, 45 (1981), 11-16, F.L. Smidth & Co., Vigerslev Alle 77, DK-2500 Valby, Copenhagen.

[6] Larsen P.M., Industrial Applications of Fuzzy Logic Control, Int.J. Man-Machine Studies, 12 (1980), 3-10.

[7] Leininger, G.G., Self-Tuning Adaptive Control of Manipulators, in: A. Danthine and M. Geradin (eds), Advanced Software in Robotics (North-Holland, Amsterdam, 1984).

[8] Mamdani E.H., Process Control Using Fuzzy Logic, Designing for Human-Computer Communication, (Academic Press, London, 1983).

[9] Mamdani E.H., Ostergaard J.J. and Lembessis E., Use of Fuzzy Logic for Implementing Rule-Based Controllers for Industrial Processes, in: Zimmermann H.J.,

Zadeh L.A. and Gaines B.R. (eds.), Fuzzy Sets and Decision Analysis, TIMS Studies in the Management Sciences, Vol 20, (North Holland, Amsterdam, 1984).

[10] Mandic N.J., Scharf E.M., Mamdani E.H., The Practical Application of a Heuristic Fuzzy Rule-Based Controller to the Dynamic Control of a Robot Arm. To be published.

[11] Procyk, T.J. and Mamdani, E.H., A Linguistic Self-Organising Process Controller, Automatica, 15 (1979).

[12] Richards, R.J., The Application of Multivariable Control to Robotics, SERC Robotics Initiative: Proceedings of the Second Grantees Conference, (Royal Holloway College, Egham, England, September 1983).

[13] Scharf, E.M., Mandic, N.J., Mamdani, E.H., A Self-Organising Algorithm for the Control of a Robot Arm, Proceedings of the Twenty-Third ISMM International Symposium: Mini and Micro-computers and their Applications, (San Antonio, Texas, U.S.A., December 1983).

[14] Scharf, E.M., Mandic, N.J., Development of Learning Algorithms for Applications in Control of Robot Arms, SERC Report GR/B 7940.0, (April 1984).

[15] Sugiyama, K., Private Communication, (November 1983).

[16] Umbers, I.G., and King, P.J., An analysis of human decision-making in cement kiln control and the implications for automation, Int. J. Man-Machine Studies 12 (1980), 11-23.

[17] Vähä, P., and Halme, A., Adaptive Digital Control for a Heavy Manipulator, in: A. Danthine and M. Geradin (eds), Advanced Software in Robotics, (North-Holland, Amsterdam, 1984).

[18] Walker, M.W. and Orin, D.E., Efficient Dynamic Computer Simulation of Robotic Mechanisms, ASME Journal on Dynamics Systems, Measurement and Control, (September 1982).

[19] Yamazaki, T., An Improved Algorithm for a Self-Organising Controller, and Its Experimental Analysis, Ph.D. Thesis, London University, (1982).

[20] Zadeh, L.A., Outline of a New Approach to the Analysis of Complex Systems and Decision Processes, IEEE Transactions SMC-3, 1 (1973), 28-44.

Industrial Applications of Fuzzy Control
M. Sugeno (ed.)

OPTIMIZING CONTROL OF A DIESEL ENGINE

Y. Murayama
Ship Research Institute
Mitaka, Tokyo
Japan

T. Terano and S. Masui
School of Engineering
Hosei University
Koganei, Tokyo
Japan

N. Akiyama
Dept. of Comm. Systems
Univ. of Electro-comm.
Chofu, Tokyo
Japan

To control a marine diesel engine minimizing its fuel consumption rate (FCR) regardless of the change of the operating conditions is practically difficult because the data of the FCR are noisy and it takes considerably long time to measure the FCR. The authors develop a new method to shorten the turnaround time required for this optimizing control by using empirical rules some of which are expressed by fuzzy numbers. During both simulator and real engine experiments, no hunting were observed and the time to reach the optimum point is reduced without sacrificing accuracy.

INTRODUCTION

The fuel consumption rate (FCR) of marine diesel engines depends on such factors as load, quality of fuel-oil, fuel injection timing, injection duration and inner pressure of the fuel pipe. The optimizing control of a diesel engine aims to keep the FCR minimum regardless of operating conditions. The "gradient method" is the most popular method for searching optimum point. But in our case, it is not so easy to detect the gradient of the FCR curve, because it is very noisy.

In this paper, the authors develop a new method of estimating the extremum point with a few number of sampled data by using empirical rules. Moreover, the wrong control operation caused by the lack of data is effectively prevented by introducing the designer's knowledge. These rules are represented linguistically with conditional statements and executed by a fuzzy optimizing controller the effectiveness of which is verified experimentally by using both a simulator and a real engine.

CONTROL SYSTEM

The controlled object is a newly developed marine diesel engine which has three vertical cylinders and is 350 PS. Its controlled variables are speed (N) and FCR (F), and the manipulated variables are fuel flow rate (Q), fuel injection timing (U), fuel injection duration (T) and inner pressure of the fuel pipe (P). In our study, the engine speed (N) is controlled by a speed governer, and T and D are fixed at constants, then FCR is optimized by adjusting U. The principle of optimizing control is as follows. FCR is measured periodically using the data of fuel flow rate and engine power. The gradient of FCR is obtained as a difference of FCR values at two different points of U. Then U is changed in proportion to the gradient. The relation between U and FCR in the steady state is shown in Fig. 1. This is almost a parabola which is suitable for applying gradient methods, but unfortunately it is disturbed by noise. The design goal is to keep FCR at the optimum value within an error of 0.2%, though the standard deviation of FCR is 1.2%.

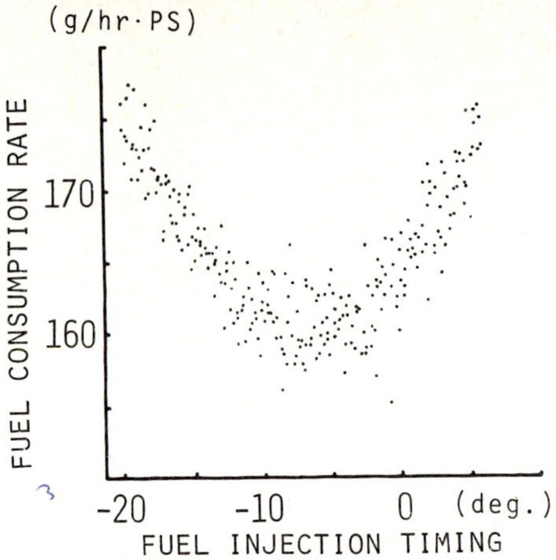

Fig. 1 Fuel Consumption Rate vs. Fuel Injection Timing

Evolutional operation [1] is one of the way for applying gradient methods to a noisy process. However, if we adopt this method, 1500 data are needed to achieve the design goal, and it takes 25 hours to search the optimum point. This is useless practically.

One of the key points of our study is how to shorten the search-time without decreasing the accuracy. The other point is how to avoid the risk of misleading the optimizing controller thereby putting the engine into a fatal condition. This risk is also caused by noise. These two problems can be solved by introducing the engineer's knowledge to the controller. Since the knowledge is subjective and qualitative, fuzzy set theory is convenient to realize it.

FUZZY OPTIMIZING CONTROL

Evolutionary operation is essentially an interval estimation of mean value. Therefore, if we reduce the number of data in order to shorten the measuring time, the confidence interval is increased, or confidence coefficient gets smaller. This is related to the misjudgement of gradient sign as shown in Fig. 2. Once misjudgement happens during descending operation of FCR, the time of arrival at optimum point is delayed remarkably. Thus the mean value of engine efficiency is decreased. For the same reason, hunting near the optimum point is also unfavorable. Steady operation without optimizing control would be better than continuous hunting.

In order to judge the sign of gradient correctly from a few data, the next empirical rule is effective.

Rule 1: FCR is measured a few times (five times in this study) at a fixed point of U. And the gradient is obtained as the difference of mean values of FCR corresponding to two different U's. If the judgement of the gradient sign is trustable, then continue descending by adjusting U in proportion to the measured gradient. If the judgement lacks confidence, then we measure five more times at the same point. If the judgement is still unreliable, then try descending.

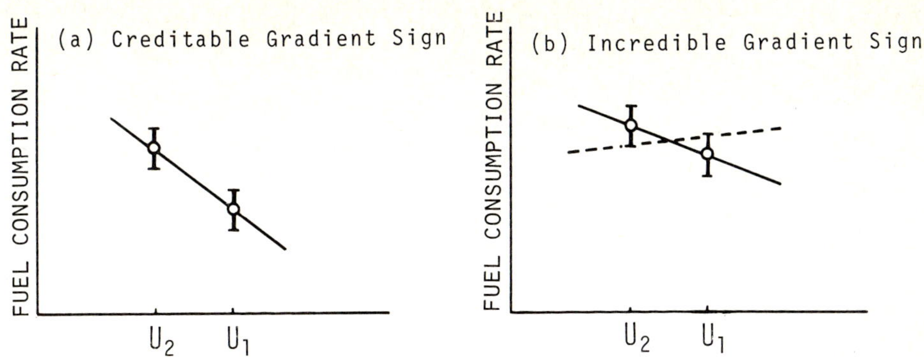

Fig. 2 Possibility of Misjudgement

Next, we consider the following rule for preventing hunting around the optimum point.

Rule 2: If the sign of gradient is changed and the number of data is more than three, then the FCR curve is identified as a parabola by using all the data through regression analysis (Fig. 3). Find the optimum point of U.

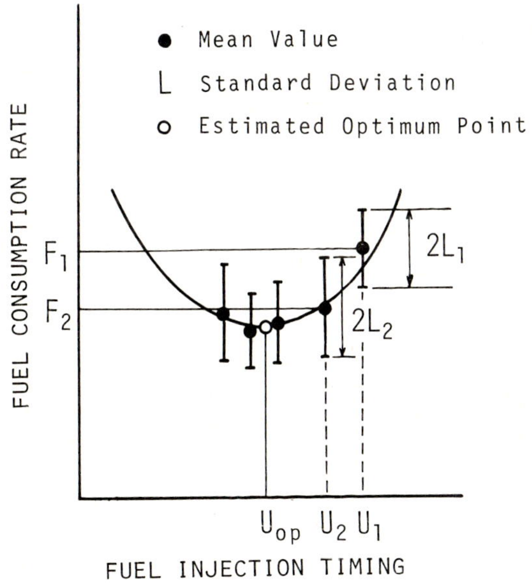

Fig. 3 Fuel Consumption Rate vs. Fuel Injection Timing

The above mentioned two rules shorten the time for finding the optimum point, but the possibility of misjudgement is also increased. Misjudgement invalidates the optimizing control, and sometimes puts engine in a fatal condition. This is avoidable by introducing the next rule.

Rule 3: If the optimum point estimated by Rule 2 is unreasonable from the view point of the designer's commonsense, repeat measuring and estimate the optimum point again. If the optimum point is regarded reasonable, then move U to that point, and also modify the designer's knowledge considering this value.

The above rules are subjective and empirical except Rule 2, therefore they are conveniently realized with a fuzzy controller. A fuzzy number is adopted to simulate Rule 1. That is, a measured value of FCR is represented with a fuzzy number of which the membership function is an isosceles triangle. The top of this triangle corresponds to the mean value of five data and its base is equal to twice the standard deviation. The gradient is obtained as the difference between two fuzzy numbers by applying the extension principle [2]. When the membership function of gradient is spread over zero, it is natural to judge its sign according to the area-ratio S of membership function as shown in Fig. 4. S is a kind

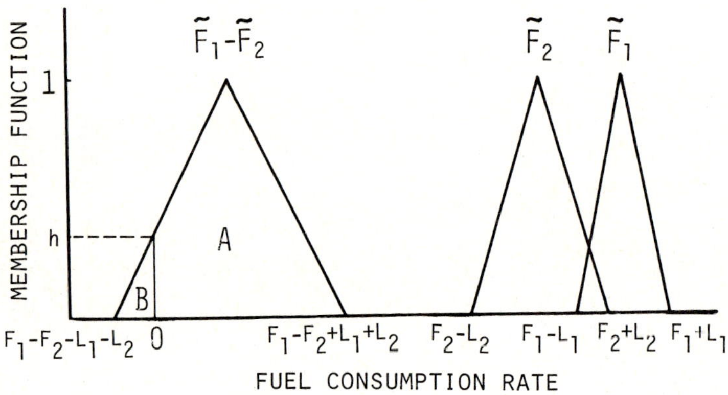

$$S = \frac{A}{A+B} = \begin{cases} 1 - h^2/2 & (F_1 \geqq F_2) \\ h^2/2 & (F_1 < F_2) \end{cases}$$

Fig. 4 Degree of Confidence of Gradient

of belief function which expresses the grade of positive gradient. It is defined as follows.

$$S = \frac{A}{A+B} = \begin{cases} 1-h2/2 & \text{for } F1 \geqq F2 \\ h2/2 & \text{for } F1 < F2 \end{cases},$$

where A, B are the positive and the negative area of the membership function respectively and h is the value of the membership function corresponding to a zero value of the gradient.

If S is larger than 0.8, then the gradient is surely positive and if smaller than

0.2, it is negative. In both cases, the controller carries out descending. When S is between 0.2 and 0.8, the gradient is uncertain, therefore measuring is continued. Rule 2 is ordinary regression analysis, so it is not necessary to explain the algorithm. As for Rule 3, we introduce a fuzzy criterion which checks that the estimated optimum point is correct. The lower and upper limits of U are -20° and 0° respectively. When U is over these limits, the engine may be broken. The designed optimum point of U at half load is given with some allowance. Therefore, the membership function of the fuzzy criterion is represented as a trapezoid whose center of upside corresponds to the design value of U. This is shown as a solid line in Fig. 5. If the value of this membership function corresponding to the estimated optimum point is larger than 0.9, then this point is considered reasonable and U is moved to this point. The criterion is revised so as to coincide with the upside center of trapezoid this point. This is shown as the dotted line in Fig. 5. If the value of the membership function is under

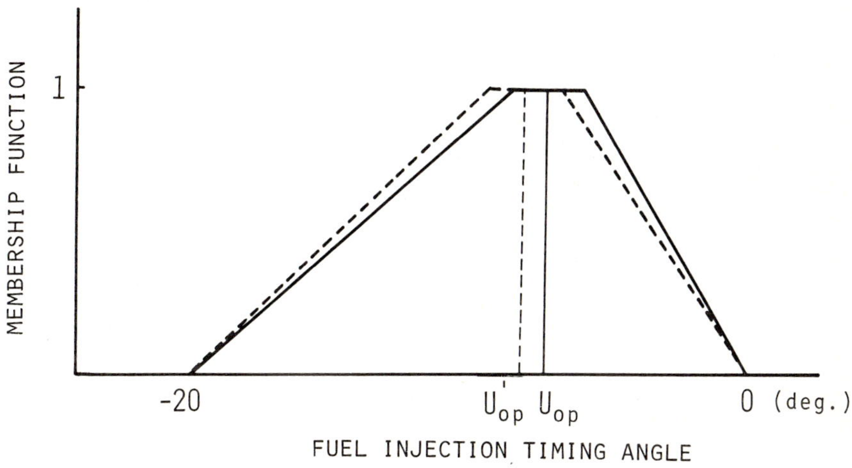

Fig. 5 Fuzzified Knowledge

0.9, then continue the descending process and estimate the optimum point again. Figure 6 shows the scheme of this fuzzy optimizing control process.

EXPERIMENTAL RESULT

In order to verify the performance of fuzzy optimizing control, both a simulator experiment and a real engine experiment were performed. It takes about 15 minutes before the real engine is settled when U is changed, and also at least one minute is needed to obtain one datum of FCR. The simulator is designed to have the same characteristics of FCR curve and noise as those of real engine. This is used to examine the performance of many different types of optimizing controller. Figure 7 shows one of the results of the simulator experiment. The solid line is the progress of U when ordinary optimizing control (evolutionary operation) is adopted and the dotted line is a result of fuzzy optimizing control. After 30 simulator experiments, the results of fuzzy control are collected in Table 1. We can see that the time to reach the optimum point in fuzzy control is about 1/30 of non-fuzzy control, and also see that there is no overshooting in fuzzy control. Among these 30 trials, the fuzzy limitation of Rule 3 functions 18 times.

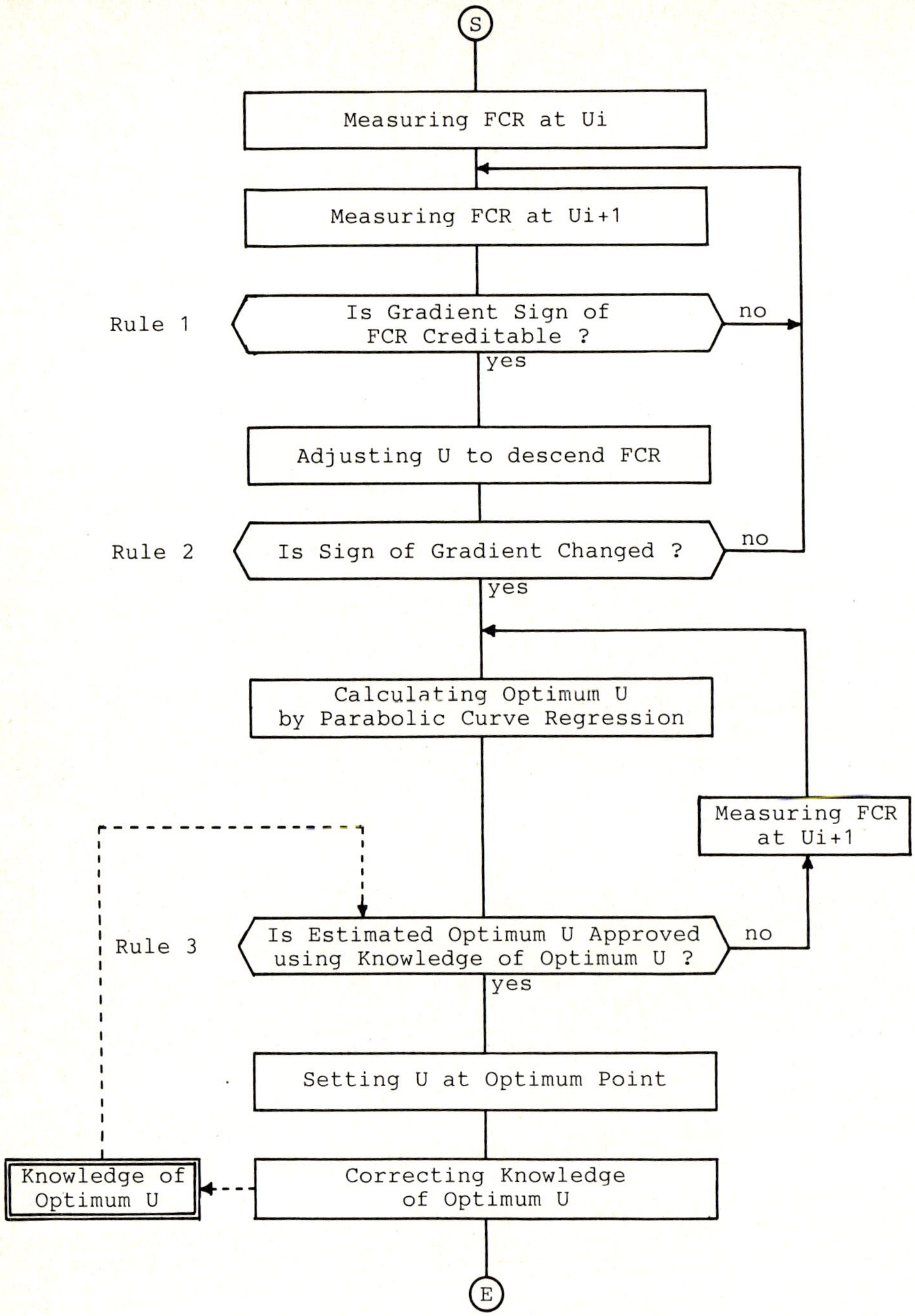

Fig. 6 Fuzzy Optimizing Control

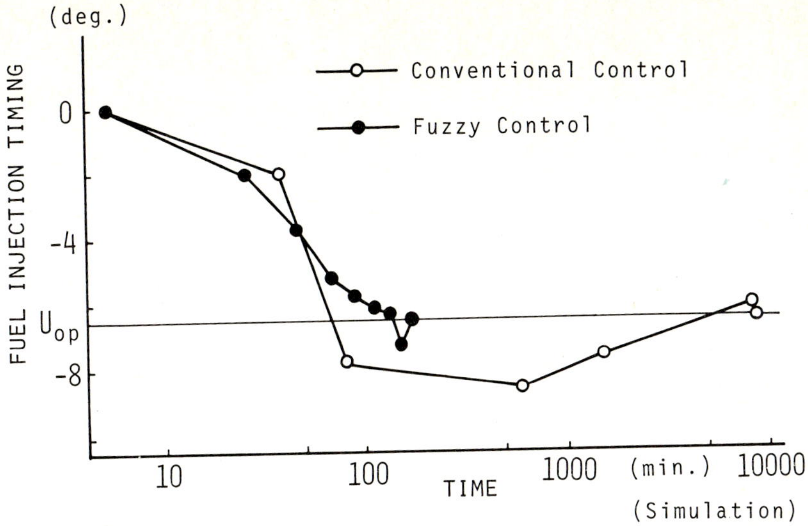

Fig. 7 Experimental Results of Optimizing Control —— Timing

Table 1 Result of Simulation

	MEAN VALUE	MAXIMUM VALUE	MINIMUM VALUE
ERROR of ESTIMATION (deg.)	-0.61	1.63	-1.32
NUMBER of STEP	5.13	9	4
NUMBER of FCR MEASUREMENT	29.2	69	20
TIME to OPTIMUM (minute)	91.2	189	65
OVERSHOOT of U (deg.)	1.25	4.70	-1.43

This means that the protection system against misoperation is indispensable for real optimizing control. Next, real engine experiments were performed. One of the results is shown in Fig. 8. Comparing Fig. 8 with Fig. 7, the overshoot is a little larger in real engine, but the time to reach the optimum point is about 100 minutes and the number of steps is about five which are almost same as the result of Table 1. The development of FCR for the real engine is shown in Fig. 9. We can see that FCR is steadily decreasing. The saving amout of FCR is shown in Table 2. The design goal is not completely attained, but it can be concluded that fuzzy optimizing control is of great utility for saving energy without risk of abnormal conditions.

CONCLUSION

It is known that the optimizing control of noisy process is very difficult. In this paper, the authors develop a fuzzy optimizing controller which shortens the time to reach the optimum point by about 1/30 over ordinary optimization, without sacrificing accuracy. Moreover, hunting around the optimum point and the risk of fatal condition are both effectively avoided. These properties of the fuzzy controller are verified by both simulation and real engine-experiments. The authors conclude that fuzzy optimizing control is of great utility and effective.

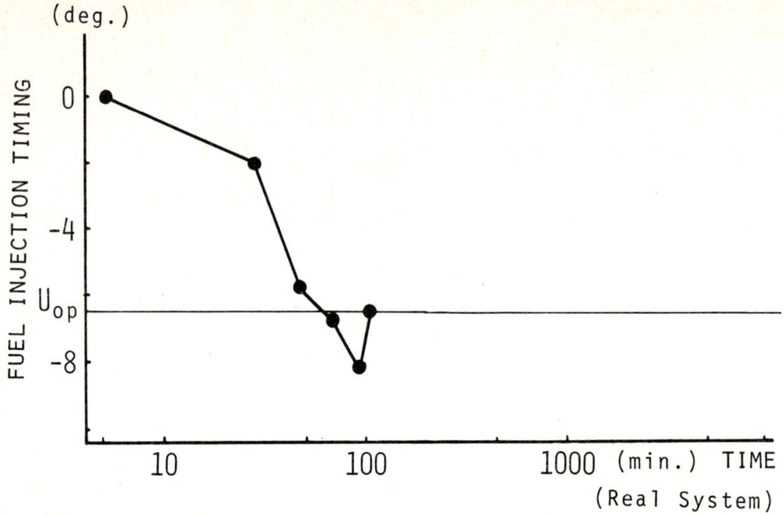

Fig. 8 Experimental Result of Fuzzy Optimizing Control — Timing

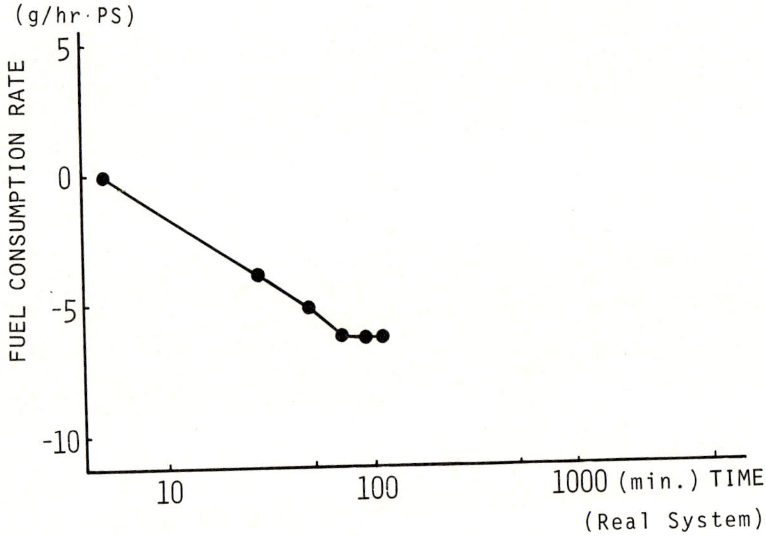

Fig. 9 Experimental Result of Fuzzy Optimizing Control — FCR

Table 2 Comparison of FCR (Real Engine)

	1/4 FULL LOAD	2/4 FULL LOAD
Without CONTROL of FCR	198.16 g/PS.hr	173.04 g/PS.hr
OPTIMIZING CONTROL of FCR	187.96 g/PS.hr	168.63 g/PS.hr

The authors are indebted to Mr. H. Fujino, who is a student of Hosei University, for his contribution to the experiments.

REFERENCES

[1] Box, G. E. P., Some general considerations in process optimization, T. ASME, (March 1960).

[2] Zadeh, L. A., The concept of a linguistic variable and its application to approximate reasoning, Inf. Sci. 8, 199; 301; 9, 43 (1975).

Industrial Applications of Fuzzy Control
M. Sugeno (ed.)

DEVELOPMENT OF PERFORMANCE ADAPTIVE FUZZY CONTROLLERS WITH APPLICATION TO CONTINUOUS CASTING PLANTS

G.Bartolini,[o] G.Casalino,[o] F.Davoli,[o] M.Mastretta,[§] R.Minciardi,[o] E.Morten[*]

[o] Dipartimento di Informatica, Sistemistica e Telematica, University of Genoa, Italy
[*] Esacontrol S.p.A., Genoa, Italy
[§] Automa Coop. r.l., Genoa, Italy

A performance adaptive fuzzy control algorithm is presented and its effectiveness is tested by means of the application to a continuous casting plant. The adaptation actions consist of modifications to the membership functions of the fuzzy subsets for input and output variables. Such modifications are taken on the basis of the overall control system's performance. The algorithm proves to give results that are comparable with those obtained by the application of self-tuning type algorithms.

1. INTRODUCTION

Control algorithms based on fuzzy sets theory have been an active and fruitful research topic during last years. More specifically, in many cases the application of such control techniques has proved to give results superior to those obtained via the application of conventional control algorithms. For a survey of both conceptual aspects and application results, the reader can refer to Tong /1/ or Gupta and Nikiforuk /2/. It is well known that the most attractive property of fuzzy control systems is their ability to emulate the behaviour of a human operator, which is based more on "vague" or "linguistic" than on quantitative criteria. For this reason, the application of fuzzy control techniques appears very useful whenever a well-defined and meaningful control objective cannot be defined, or the system to be controlled cannot be modelled in a simple way.

Nevertheless, it is well recognized that, besides the ability of taking decisions, or developing control rules following vague and subjective specifications, a second fundamental feature of a human operator's behaviour is learning, that is, the ability to correct his operating rules and to make use of his own experience.

A former version of this paper was presented at the Sixth European Meeting on Cybernetics and Systems Research, Vienna, Austria, April 13-16, 1882.

Unfortunately, most of the existing literature dealing with fuzzy control, stresses the first aspect, with the only valuable exception (in the authors' knowledge) of the work of Procyk and Mamdani /3/, who have developed a self-organizing process controller based on a linguistic description of the control strategy.

One of the purposes of this paper is to explore the possibilities of inserting learning characteristics into control algorithms based on fuzzy sets theory. In this connection, the approach that will be described in the paper represents an attempt to extend the philosophy of performance adaptive control systems (see Saridis /4/) to the class of fuzzy control algorithms. Essentially, the method we propose is that of modifying, based on the overall control system performance, the fuzzy specifications leading to the synthesis of the control algorithm. In this way, it is possible to obtain control systems that become automatically "tuned", in case of unknown or complex processes, or that are able to cope with time-varying effects in the process.

A case study is considered in the paper, relevant to the control of a continuous casting plant. The control problem that is dealt with seems to be particularly suited for the application of the above described techniques, due to the complexity of the plant and to the vagueness of the design specifications. A comparison of the performance obtained by means of the proposed fuzzy controller with those given by conventional PID regulator or with adaptive control is presented in the paper, and the effectiveness of the proposed techniques is shown.

The paper is structured in the following way. In the next section, a short review of fuzzy control systems is given, and in section 3 a brief description of the plant is presented. Section 4 contains the development of the "learning" fuzzy control algorithm which is the original contribution of our work, while in section 5 a performance analysis of the proposed algorithm is detailed, together with a comparison with different control techniques. The results obtained show that the method can have performances comparable with those of self-tuning type algorithms.

2. A SHORT REVIEW OF FUZZY CONTROL SYSTEMS

For the convenience of the reader, in this section we shall summarize the basic theory of fuzzy control systems, following essentially Kickert and Van Nauta Lemke /5/ and Tong /1/.

The fuzzy control system derives from a set of fuzzy implications of the type "if A then B", where A and B are fuzzy subsets on the universes of discourse of output and input, respectively. Consider, for simplicity, the case of a single input (u) - single output (y) system and suppose there are I of such fuzzy implications. Each of these rules associates a fuzzy input subset to a fuzzy output subset, represented by their membership functions:

$$\mu_i(y) \longrightarrow \nu_i(u) \qquad i=1,2,\ldots,I \qquad (1)$$

These fuzzy implications represent a sort of "qualitative" feedback relationship between the output and the input of the process, and can be regarded as an attempt at modelling the behaviour of a human operator, who chooses his input action after observation of the output.

Clearly, the actual input action u_o must be determined by the control system as a deterministic function of the measured output y_o. To this end, the following operations are performed. First, the membership values for the output y_o are computed for each rule

$$\mu_i(y_o) \qquad i=1,2,\ldots,I$$

Then, the implied fuzzy subset for input, corresponding to the i-th rule, has the membership function

$$\lambda_i(u) = \min\left[\mu_i(y_o);\quad \nu_i(u)\right] \qquad (2)$$

and the overall fuzzy subset for input is obtained by the union operator

$$\lambda(u) = \max_{i=1,\ldots,I} \min\left[\mu_i(y_o);\quad \nu_i(u)\right] \qquad (3)$$

$\lambda(u)$ is a fuzzy subset that ranges over all values of the output. Finally, the actual input action u_o is chosen as that maximizing the membership function of this fuzzy subset, that is,

$$\lambda(u_o) = \max_{u} \max_{i=1,\ldots,I} \min\left[\mu_i(y_o);\quad \nu_i(u)\right] \qquad (4)$$

Apparently, the above outlined derivation of the input u_o, after observation y_o, does not need to be performed on-line; obviously, a look-up table can be formed, which associates input actions to observed output values. This table can be regarded as a deterministic feedback controller, and the above described procedure represents the synthesis algorithm based on the set of the I fuzzy rules. Clearly, a discretization of possible values of input and output must take place.

A straightforward generalization of the above concepts allows to consider the case of multi-output and multi-input systems. For instance, in case of two outputs, y^1 and y^2, and one input, u, the membership function (2) is replaced by

$$\lambda_i(u) = \min\left[\mu_i^1(y_o^1);\quad \mu_i^2(y_o^2);\quad \nu_i(u)\right] \qquad (5)$$

where the symbols have an obvious significance, which corresponds to the implied fuzzy subset relevant to the i-th fuzzy rule

$$\mu_i^1(y^1), \quad \mu_i^2(y^2) \longrightarrow \nu_i(u) \tag{6}$$

Then, the derivation of the actual input action follows exactly the same procedure as above. In case of two inputs, the algorithm to find the input actions is simply duplicated, and so on.

The key point of the overall procedure to derive the deterministic input-output mapping (termed, with a little abuse, fuzzy controller), is, of course, the definition of the fuzzy rules and, more specifically, of the membership functions of the fuzzy input and output subsets associated by each rule. One can object that to define these rules is intrinsically as difficult as to establish the performance index in an optimal control framework, so that the use of the above outlined approach simply shifts the problem of defining the performance measure of the control system. The answer to this objection is primarily that the fuzzy rules can be deduced from experience, that is, by observation of the human operator's behaviour. Besides, the scope of the paper is to show that simple algorithms can be devised for automatically adjusting the shape of the primitive membership functions, based on the performance obtained by the control system, so that the initial guess about the fuzzy rules turns out to be not extremely critical.

3. A BRIEF DESCRIPTION OF THE PLANT

Fig. 1 shows a schematic picture of a continuous casting plant, where the three subsystems which constitute the whole plant, the ladle, the tundish and the mould, are represented. f_l, f_t and f_m are the output metal flows from the ladle, the tundish and the mould, respectively. f_t is controlled by adjusting the position of a stopper rod, while f_l is controlled by adjusting a sliding nozzle valve. Instead of considering the overall multivariable control problem for the whole plant, this paper deals only with a subproblem, namely that of regulating the metal level in the tundish h, by adjusting the sliding nozzle valve. Any interaction with other subsystems of the plant is regarded as a disturbance. The main reason for this restriction is that the overall control problem turns out to be, at a first approach, considerably complex. Thus, it seems reasonable, at least in our introductory study, to limit ourselves to consider a single-input single-output control problem.

At the present time, the control function in continuous casting plants is mostly accomplished by human operators. Only recently, automatic control systems based on PID regulators have been introduced. In a former work, the authors /6/ have proposed an adaptive control approach which has proved to be quite effective and to give superior performance with respect to conventional controllers.

The control objective of the problem considered in the paper is that of keeping the metal level in the tundish as constant as possible. The reason for this is twofold: i) to avoid impurities being sucked into the mould; ii) to keep the metal output flow from the tundish,

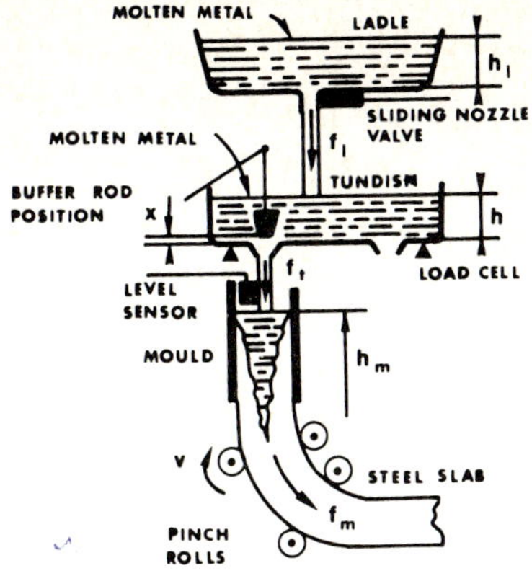

Fig. 1. A schematic picture of the plant.

f_t, as constant as possible. The control action is simply a command which can increase or decrease the metal input flow into the tundish. This command is allowed to assume only three different values: i) open (this corresponds to a forward fixed displacement of the sliding nozzle which increases the opening of the input valve); ii) close (this corresponds to a backward fixed displacement of the sliding nozzle which decreases the opening of the input valve); iii) no movement of the sliding nozzle. One of the requirements of the problem is that of minimizing the number of variations in the motion condition of the sliding nozzle (due to the particularly "hard" conditions in which the sliding nozzle valve operates). That, of course, can be in conflict with the objective of regulating the level h.

Several noise sources and many nonlinear and time varying effects (for instance, the emptying of the ladle and wear, tear and encrusting phenomena over the sliding nozzle valve) are present in the system. All these facts, and the difficulty of synthesizing an optimal control law which accounts for the physical complexity of the overall system, make an adaptive control approach attractive /6/. An alternative to this approach can be that of fuzzy control algorithms, whose application to the above described plant is reported in this paper.

4. DEVELOPMENT OF A PERFORMANCE ADAPTIVE CONTROL ALGORITHM

In this section a procedure will be developed which will lead to a fuzzy control algorithm characterized by a certain degree of adaptivity, on the basis of the obtained performance. Such an algorithm, as already pointed out, can be also viewed as a means for automatically "tuning" a fuzzy controller, without being involved into a deep analysis of the plant and into tedious efforts of reproducing the human operator's behaviour.

The basic scheme of the performance adaptive fuzzy controller (hereafter PAFC) is that of Fig. 2. Clearly, the key role is played by the performance evaluation and adaptation blocks. In the following, we shall detail the structure of these blocks for the particular case study dealt with in the paper. Nevertheless, it will be apparent that similar considerations could be made in general. A fuzzy controller for the above introduced control problem will be considered, with two inputs (that is, plant outputs), namely h (metal level in the tundish) and dh (time derivative of h), and one output (plant input) u, that is, the action over the sliding nozzle valve. A set of 12 fuzzy rules has been chosen, corresponding to Table I. The membership functions of the fuzzy subsets for h and dh have been chosen to be of the following type

$$f_1(x) = (1 + (a\ (x-c))^b)^{-1} \tag{7}$$

$$f_2(x) = e^{-(x-m)^2/2\sigma^2} \tag{8}$$

More specifically, the former is associated with the fuzzy subsets "high" and "low" for h, and with the fuzzy subsets "positive", "negative", "very positive" and "very negative" for dh, with suitable choices of the parameters a, b, c. The latter, instead, is associated only with the subset "medium" for h. In Fig. 3 a choice

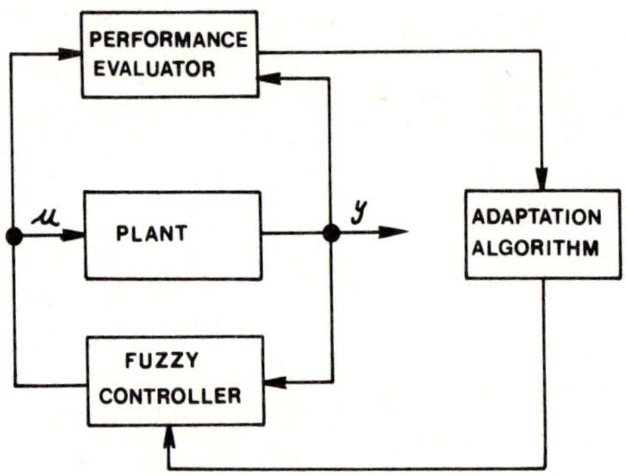

Fig. 2. The basic scheme of the performance adaptive fuzzy controller.

Table I. Fuzzy rules of the control algorithm

1) If h "low" and dh "very negative", then u "high".
2) If h "low" and dh "negative", then u "high".
3) If h "low" and dh "positive", then u "high".
4) If h "low" and dh "very positive", then u "medium high".
5) If h "medium" and dh "very negative", then u "high".
6) If h "medium" and dh "negative", then u "medium high".
7) If h "medium" and dh "positive", then u "medium low".
8) If h "medium" and dh "very positive", then u "low".
9) If h "high" and dh "very negative", then u "medium low".
10) If h "high" and dh "negative", then u "low".
11) If h "high" and dh "positive", then u "low".
12) If h "high" and dh "very positive", then u "low".

of the membership functions for each fuzzy subset is shown. As previously mentioned, a discretization of the values of h and dh is necessary. We have chosen to consider 21 possible discrete values for h and 15 possible discrete values of dh. Besides, when the values of h and dh are recognized to be out of the ranges corresponding to the membership functions of Fig. 3 (that is, for instance, when h > 85 cm), then we consider to be out of the normal operating conditions,

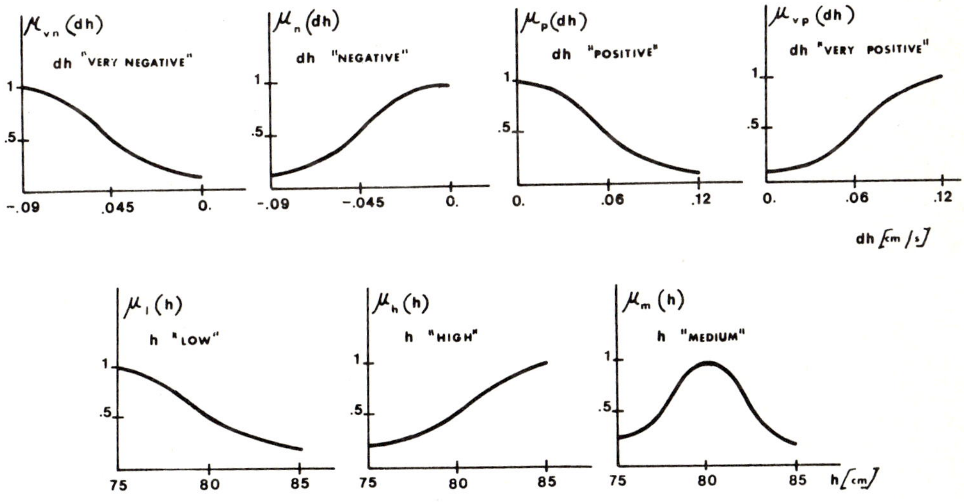

Fig. 3. A choice of the membership functions of the fuzzy subsets for h and dh.

and a movement command of the sliding nozzle is given.

We have already pointed out that the control variable u is discretized into three values, namely, 1 (open), -1 (close), and 0 (no movement). Thus, the membership functions of the fuzzy subsets for the input are simply represented by three values. A possible choice of these membership functions is shown in Fig. 4. The following performance indexes have been adopted:

(1) the average square error in the controlled level (with respect to the specified set-point h_{sp}=80 cm), $\overline{e^2}$;

(2) the average error in the controlled level, $\bar{e}$;

(3) the average absolute error in the controlled level, $\overline{|e|}$;

(4) the maximum of the absolute error in the controlled level, $|e|_{max}$;

(5) the number n_1 of consecutive variations in the command to the sliding nozzle;

(6) the number n_2 of variations, in the command to the sliding nozzle, during a certain time interval.

The primary objective of the control system is that of maintaining the metal level in the tundish as close as possible to the given set-point. Thus, the adaptation mechanism is first of all devoted to the attainment of a good performance as regards the control of the level. Only secondarily the objective of reducing the number of command variations is sought.

The performance indexes 1), 2) and 4) are all evaluated over a fixed observation period. The block diagram of Fig. 5 shows the procedure for generating possible adaptation actions. α and β are specified threshold values (actually, the choice of these parameters may have a great effect on the overall control system performance). It

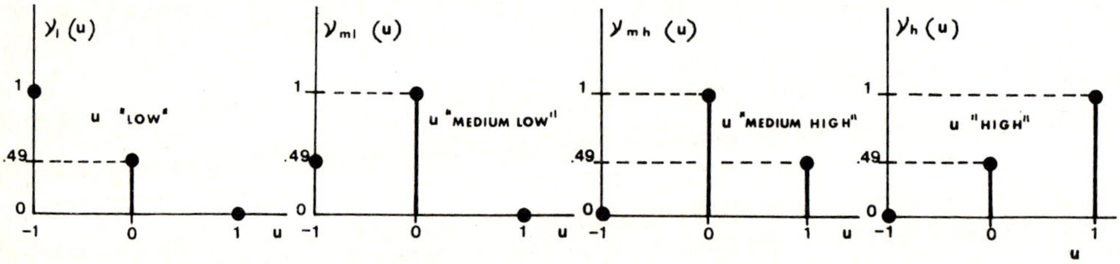

Fig. 4. A choice of the membership functions of the fuzzy subset for u.

is worth remarking that the choice of the observation period constitutes another critical point; clearly, the longer this interval, the more significant the evaluated performance indexes $\bar{e}$, $\overline{e^2}$, $\overline{|e|}$, but also the poorer the control system's readiness characteristics. The adaptation action a) is taken in a situation in which the metal level is "too low"; on the contrary, the adaptation action b) must face a situation in which the metal level is "too high". The adaptation action c) corresponds to a situation in which there is a "lack of precision", and finally, the adaptation action d) corresponds to a situation in which a "too high error" in the metal level is recognized, even for a single instant.

While the procedure for generating possible adjustments of the control system, corresponding to the block diagram of Fig. 5, is performed at a constant rate, a second adaptation procedure is performed at every sampling interval. This procedure is devoted to the generation of possible adaptation actions having the objective of reducing the number of command variations, and corresponds to the block diagram of Fig. 6. The higher frequency is due to the

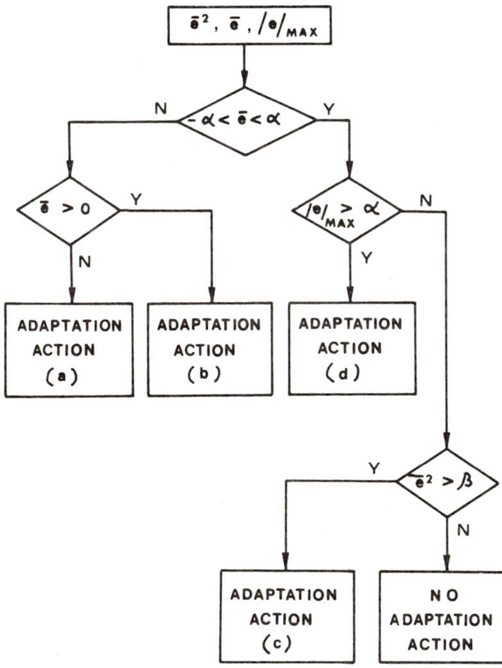

Fig. 5. The procedure for generating possible adaptation actions having the objective of improving the control of the metal level.

necessity of promptly recognizing a possible too high number of command variations. The indexes n_1, n_2, and $|\bar{e}|$ are evaluated over a finite past time horizon, and are updated at each sampling period. Note that, when an excessive number of command variations is recognized, but this fact can be ascribed to the attempt of reaching a satisfactory control of the level, then no adaptation action occurs, at that sampling instant. On the contrary, the adaptation action e) takes place whenever a satisfactory level behaviour is recognized and an excessive number of command variations cannot be justified. Clearly, also ε_1, ε_2 and γ are parameters that must be carefully chosen. Let us now briefly explain the various adaptation actions in detail. The adaptation action a) corresponds to modifications of the membership functions of the fuzzy subsets for variables h, dh, u, which are shown in Fig. 7 (of course, every modification implies the recomputation of the controller look-up table). The reasons for such modifications are evident. For instance, as to the h variable, the degree to which certain values of h are recognized as "low" is incremented, while the corresponding degrees for "medium" and "high" are decremented: this, of course, has the objective of improving the "ability" of recognizing low values of h. Similar considerations could be made as regards the other membership functions modifications. The adaptation action b) is exactly symmetrical to action a). For the sake of brevity we can condense the adaptation actions c) and d) into a single action which improves the sensitivity . A

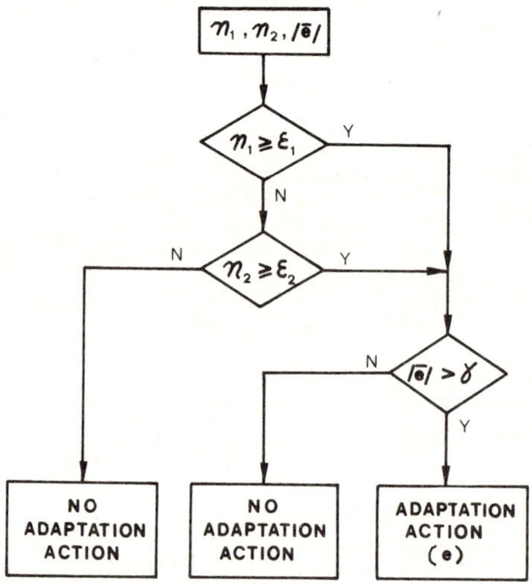

Fig. 6. The procedure for generating possible adaptation actions having the objective of reducing the number of command variations.

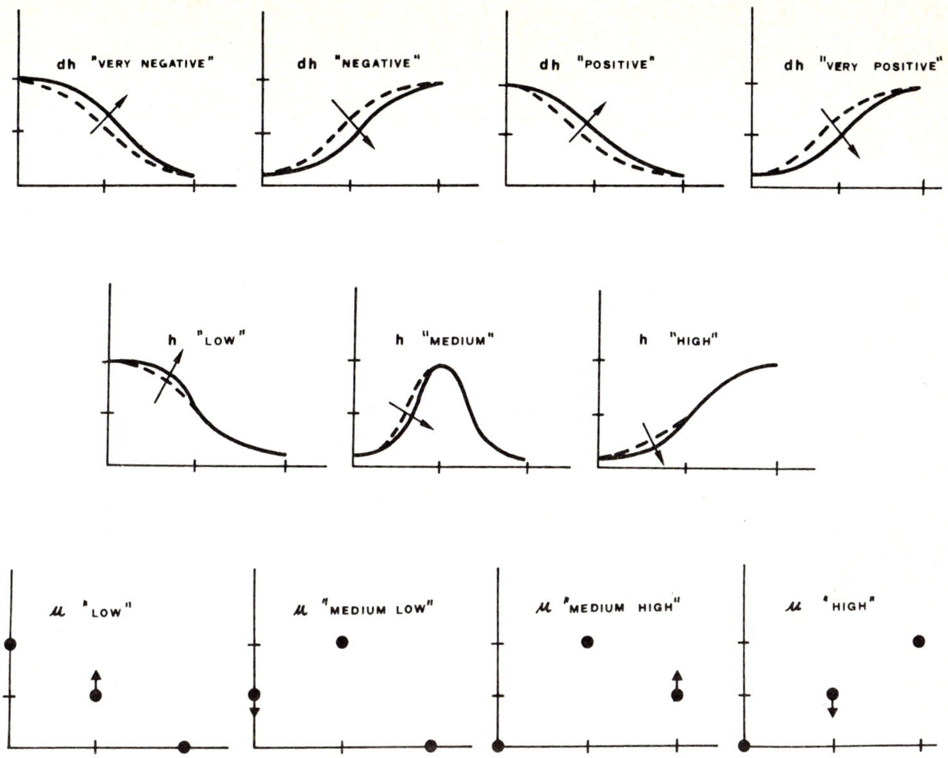

Fig. 7. Modifications of the membership functions of the fuzzy subsets for variables h, dh, u, corresponding to the adaptation action a).

sample of the corresponding modifications to the membership functions is represented in Fig. 8. On the contrary, the adaptation action e) consists of a sensitivity reduction (with the objective of minimizing the command variations), and thus turns out to be exactly symmetrical to the action corresponding to the modifications due to actions c) and d). We do not go into quantitative details as regards the modifications of these membership functions: generally speaking, a strategy has been adopted which corresponds to correcting the shapes of the membership functions proportionally to the undesired recognized effects which are the reasons for the adaptation. Clearly, a key design problem is that of establishing adjustment strategies capable of giving the necessary adaptation quickness, without causing instability or oscillations of the control

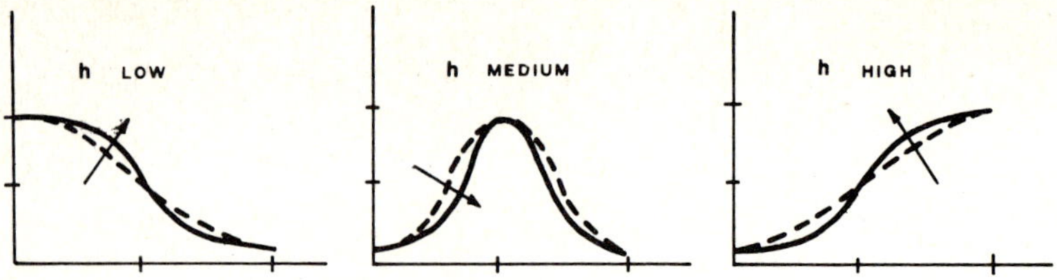

Fig. 8. A sample of the modifications of the membership functions, corresponding to actions c) and d).

law. Of course, the above described adaptation mechanism is not unique, nor the best possible. Actually, many diverse possibilities exist: for instance, one might choose not only to modify the existing fuzzy rules, but even to add new fuzzy rules, in some special situation. The scope of this section has been only to show how, with a little ingenuity, performance adaptation schemes can be found even for fuzzy control systems.

5. RESULTS OF THE APPLICATION OF THE PERFORMANCE ADAPTIVE CONTROL ALGORITHM

The application of the control algorithm introduced in the previous section has been tested by means of simulations. To this end, a simulation code of the plant has been developed which accounts all the complexity of the controlled process.

A first test consisted in the application of the PAFC algorithm with the following parameter values: $\alpha = 2.5$ cm, $\beta = 1$ cm^2, $\gamma = 1$ cm, $\varepsilon_1 = 4$, $\varepsilon_2 = 6$. The whole simulation period lasted 40 minutes, and the observation period was taken to be 2 minutes. In Fig. 9 the behaviour of the output variable h is plotted. The number of command variations was 43. As to the influence of the choice of the various parameters, several tests have been performed. For instance, the effect of variations of parameter β is summarized in Table 2 (all other parameter values are as above). As is consistent with intuition, increasing values of β yield worse level behaviour, but lower numbers of command variations.

We do not report further results for the sake of brevity. However, it is worth remarking that in any case, the presence of the adaptation function has proven to be essential to obtain a good performance of the control system.

At this point, it is quite natural to compare the PAFC algorithm

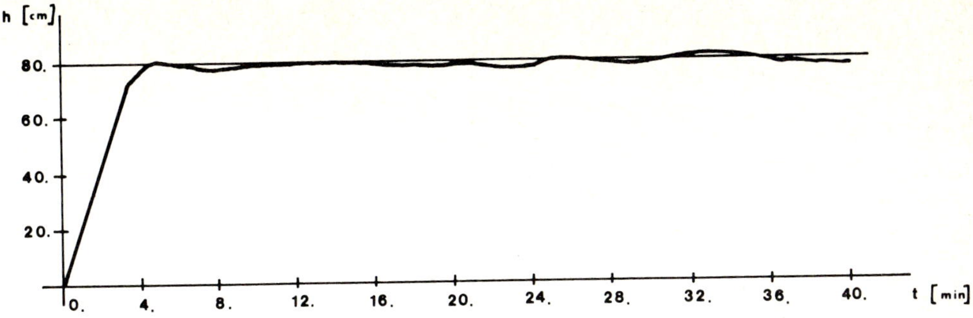

Fig. 9. Behaviour of the output variable h.

Table II. Effect of variations of parameter β

β [cm^2]	$\overline{e^2}$ [cm^2]	number of command variations
0.1	0.29	92
0.25	0.48	56
0.75	0.43	47
1.	0.71	43
2.	1.15	38

with the adaptive control scheme developed in /6/, which combines Least Squares identification with the minimization, with respect to u, of the cost

$$E\left\{ (h_{i+k+1} - h_{sp})^2 + p\,|u_i - u_{i-1}| \right\} \tag{9}$$

where k is the system's fixed delay and p is a weighting coefficient. Some of the results obtained are summarized in Table III, for different values of p. Apparently, an inspection of Tables II and III shows that the adaptive control and the PAFC schemes yield comparable results. A slight advantage of the adaptive control approach is balanced by the fact that a careful choice of the parameters of the PAFC scheme allows to obtain better results than those of Table II. For instance, the following parameter values: α = 2.5 cm, β = 0.5 cm^2, γ = 0.707 cm, ε_1 = 4, ε_2 = 6, give $\overline{e^2}$ = 0.33 cm^2 and 43 command variations. We can also conclude that the PAFC algorithm is certainly superior to a conventional PID controller, since the latter was shown to yield considerably poorer results than those obtained by means of adaptive control /6/.

Table III. Performances obtained by means of the adaptive control scheme.

p	$\overline{e^2}$ [cm^2]	number of command variations
$0.2\cdot10^{-5}$	0.038	103
$0.6\cdot10^{-5}$	0.324	65
$0.1\cdot10^{-4}$	0.5	49

REFERENCES

/1/ Tong, R.M., A control engineering review of fuzzy systems, Automatica 13 (1977) 559-569.

/2/ Gupta, M.M. and Nikiforuk, P.N., Feedback control of industrial processes via fuzzy set theory: some remarks, Proc. JACC (1978) 37-52.

/3/ Procyk, T.J. and Mamdani, E.H., A linguistic self-organizing process controller, Automatica 15 (1979) 15-30.

/4/ Saridis, G.N., Towards the realization of intelligent controls, Proc. IEEE 67 (1979) 1115-1133.

/5/ Kickert, W.J.M. and Van Nauta Lemke, H.R., Application of a fuzzy controller in a warm water plant, Automatica 12 (1976) 301-308.

/6/ Bartolini, G., Casalino, G., Davoli, F., Mastretta, M., Minciardi, R. and Morten, E., Development and application of adaptive control methods in continuous casting plants, Proc. 4th Int. Conf. on Trends in On-line Computer Control Systems, Coventry (1982), IEE Conf. Publ. 208, 34-37.

Industrial Applications of Fuzzy Control
M. Sugeno (ed.)

A FUZZY LOGIC CONTROLLER FOR AIRCRAFT FLIGHT CONTROL

Lawrence I. Larkin

Corporate Engineering & Manufacturing
General Electric Company
Bridgeport, CT 06601

ABSTRACT

This paper describes a model of an autopilot controller based on fuzzy algorithms. The controller maneuvers an aircraft from level flight into a final-approach flight path and maintains the aircraft along the glide path until just before touchdown. To evaluate the performance and effectiveness of the model, the aircraft response to controller actions is simulated using flight simulation techniques. The effect of varying parameters and different defuzzification strategies on controller performance is analyzed.

INTRODUCTION

In aircraft flight the final approach and landing are critical and require constant monitoring and control. To reach the runway, the pilot must "aim" the flight path of the aircraft toward the runway and must follow a steady course directly into the landing area. A direction-finding Instrument Landing System (ILS) provides a navigational radio signal to guide approaching aircraft to the runway. Available in most airports, the ILS transmits two separate orthogonal radio beams whose intersection forms a signal that the aircraft follows all the way to the airport (Fig. 1). At the far end of

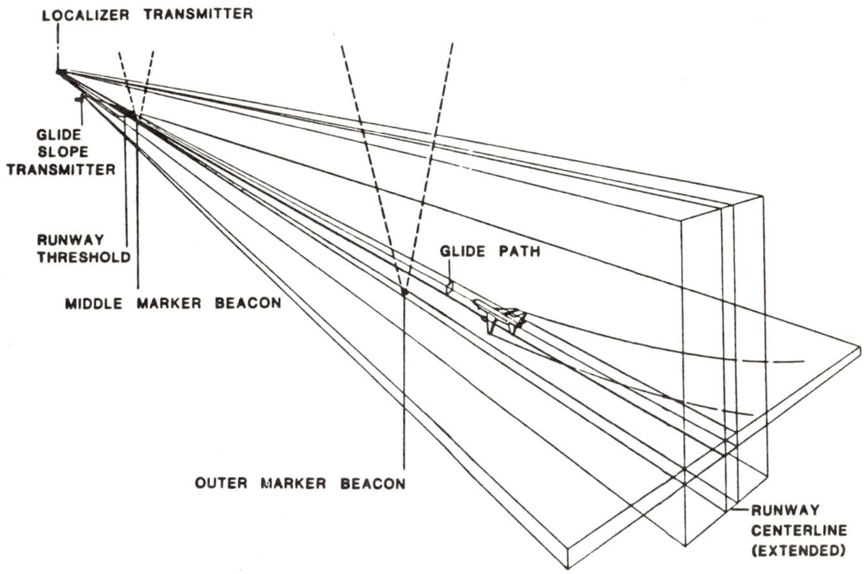

Fig. 1. The Instrument Landing System (ILS).
(Adapted from Ref. 2)

the airport runway, the "localizer" transmitter emits a beam along a plane vertical to the runway. Near the opposite end of the runway, approximately 150 yards from the edge, the "glide-slope" transmitter emits a beam along a plane approximately 2.5 to 3.0 degrees from the horizontal. The path formed by the intersection of these two beams, referred to as the "glide path," provides the optimum landing trajectory.

Typically, an aircraft intercepts the ILS signal about eight miles from the runway. A cockpit instrument, known as the ILS Glide-Path Indicator, indicates the position of the aircraft relative to the ILS signal and is used by the pilot to maintain the aircraft along the ILS trajectory. Any deviations from the prescribed glide path will cause the aircraft to either miss the runway or land at an excessive speed (which may cause the aircraft to overrun the runway). Throughout the entire approach the pilot must maintain a straight course at a relatively constant rate of descent and speed. In general, the course corrections required are small. The pilot determines the extent of the deviation from the glide path from instrument readings and makes the corrections by adjusting the aircraft controls (throttle, elevators, flaps, etc.).

The course corrections are achieved by a process of successive approximation. After determining from instrument readings what the deviation from course is, the pilot makes a decision on which control(s) to adjust, makes a mental estimate of how much of an adjustment is required, and then makes the adjustment. The pilot then looks again at the instruments to determine if the adjustment has corrected the course deviation (error). The pilot may have under- or over-compensated for the error and hence must make additional adjustments — each one being progressively finer — to place the aircraft back on course. The specification of control actions is inherently imprecise. That is, the pilot characterizes adjustments as "giving it a lot of throttle" or "increasing the pitch a little."

The aircraft control process is complicated by the fact that there is a large degree of coupling between the various aircraft controls. For example, an increase in the pitch angle will cause a decrease in airspeed. Furthermore, the same control adjustments will yield totally different results depending on the flight conditions. For example, at high airspeeds, an increase in pitch angle will result in an upward acceleration — causing the aircraft to climb and lose some groundspeed. At low airspeeds, however, the same action will cause the aircraft to drop and lose speed. In this study, a model of an autopilot controller based on fuzzy algorithms is developed. The controller maneuvers the aircraft from level flight into a final-approach flight path and maintains the aircraft on the ILS course until just before touchdown, at which time the pilot takes over control. To evaluate the performance and effectiveness of the model, aircraft flight under the control of the fuzzy-logic controller is simulated by computer model.

THE SYSTEM

The system has two major components: the controller and the flight simulator (aircraft) that interact with each other and form a closed-loop control system (Fig. 2). To achieve the desired glide-slope trajectory, the controller issues flight commands (e.g. increase speed, decrease pitch angle) to the simulator which, in turn, computes new flight conditions (e.g. airspeed, rate of descent). The controller uses this information to generate new flight commands that will maneuver the aircraft into the desired trajectory.

Controller

The controller has three inputs: rate of aircraft descent (fpm), deviation from desired speed (mph), deviation from glide-slope trajectory (degrees), and two outputs: engine speed change (rpm) and elevator angle change (degrees). The control system itself consists of three stages: fuzzification, decision logic and defuzzification. The fuzzification stage converts real-number input values into fuzzy values. The decision logic stage processes the input data and computes the controller outputs. These outputs, which are fuzzy values, are converted into real numbers by the defuzzification stage.

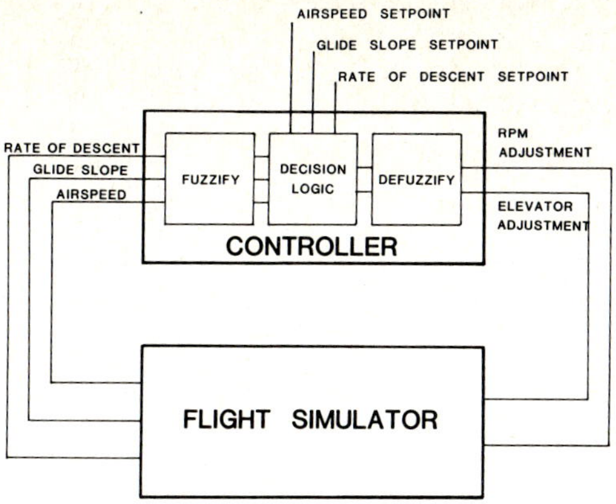

Fig. 2. Block Diagram of System.

Fuzzification Stage

The fuzzification stage converts a real number into a fuzzy singleton within a Universe of Discourse U. The Universe U is quantized to 21 levels ranging in value from -10 (largest negative magnitude) to +10 (largest positive magnitude) and encompasses five fuzzy subsets, termed NB (Negative Big), NM (Negative Medium), IC (Insignificant Change), PM (Positive Medium) and PB (Positive Big) (Fig. 3).

U	NB	NM	IC	PM	PB
-10	1.000	0.	0.	0.	0.
-9	0.920	0.080	0.	0.	0.
-8	0.720	0.320	0.	0.	0.
-7	0.320	0.680	0.	0.	0.
-6	0.080	0.920	0.	0.	0.
-5	0.	1.000	0.	0.	0.
-4	0.	0.920	0.080	0.	0.
-3	0.	0.680	0.320	0.	0.
-2	0.	0.320	0.680	0.	0.
-1	0.	0.080	0.920	0.	0.
0	0.	0.	1.000	0.	0.
1	0.	0.	0.920	0.080	0.
2	0.	0.	0.680	0.320	0.
3	0.	0.	0.320	0.680	0.
4	0.	0.	0.080	0.920	0.
5	0.	0.	0.	1.000	0.
6	0.	0.	0.	0.920	0.080
7	0.	0.	0.	0.680	0.320
8	0.	0.	0.	0.320	0.720
9	0.	0.	0.	0.080	0.920
10	0.	0.	0.	0.	1.000

Fig. 3. Universe of Discourse U and Corresponding Membership Functions for NB, NM, IC, PM and PB Fuzzy Subsets.

The five fuzzy subsets and their membership functions were subjectively assigned to define, in a quantitative manner, the meaning of qualitative terms such as Positive Medium. All of the fuzzy values in this study are subsets of U, i.e., the Universe of Discourse is the same for all fuzzy variables.

By definition, a fuzzy singleton is a fuzzy set S whose support is a single point in U and is represented as

$$S = \mu/x \tag{1}$$

where μ is the grade of membership of x in S. Given a real number R, the corresponding fuzzy singleton in U is given by

$$S_R = \mu_R/x_R \tag{2}$$

where μ is the grade of membership of x_R in S_R and

$$x_R = \text{ENTIER}(R/d) \tag{3}$$

where d is a constant representing the scale of the quantization levels in U. In the case of airspeed, for example, each increment (quantization level) in U represents 5 mph; thus d is equal to five.

Decision Logic Stage

The decision logic stage consists of a set of 125 decision rules and associated logic that determine the values of the control variables for a given set of input values. The rules were formulated in linguistic rather than mathematical terms. The actual rule set was generated by interrogating an experienced pilot and asking him a number of highly structured questions. The pilot was given an objective: to land the aircraft while maintaining an airspeed of 75 mph, rate of descent of 350 fpm and glide-slope angle of three degrees (Any other set of values could have been used; the parameters used here are applicable to a small single-engine aircraft). He was then asked what control actions he would take given a stipulated set of flight conditions. All questions had the same format, namely:

> "If the current rate of descent is around x fpm, the current airspeed is around y mph above/below the desired airspeed and the glide slope is about z degrees above/below the desired glide slope, by about how much would you change the engine speed and the elevator angle?"

The variables x,y and z were each assigned five different values, each value representing a positive or negative deviation (error) from the objective value. For example, the glide-slope error was varied from +2 to -2 degrees in one degree increments. The above question was asked for every possible combination of values for x,y and z. The pilot's response was limited to five choices:

- increase substantially
- increase slightly
- no change
- decrease slightly
- decrease substantially

where, for example, an "increase rpm slightly" response was taken to mean an increase of "about 100 rpm."

Each rule was represented linguistically as a compound IF-THEN statement namely:

IF rate_of_descent = A, airspeed = B and glide_slope = C (4)

THEN rpm_change = X, elevator_angle_change = Y

where A,B,C are the values of the input variables and X,Y those of the output variables. These variables were implemented as fuzzy values in the Universe of Discourse U. The five fuzzy subsets —NB, NM, IC, PM and PB — represent the possible values A,B,C,X and Y can individually assume. These five fuzzy subsets directly correspond to the five possible choices the pilot was given for his responses.

The 125 rules represent all possible combinations of the three input variables (rate_of_descent, airspeed, glide_slope) and their respective values (NB, NM, IC, PM, PB). A typical rule thus looks like:

IF rate_of_descent = PM, airspeed = NB and glide_slope = PB (5)

THEN rpm_change = PM, elevator_angle_change = IC.

In actual practice, the set of rules was encoded and implemented as a look-up table. Each row in the 125-entry table represents one rule. Within each row there are five "cells." Each cell contains the fuzzy subset representing the value of the respective input/output variable for the given rule. The tabular representation of the rule shown in Eq. (5) is depicted in Fig. 4. The fuzzy sets were

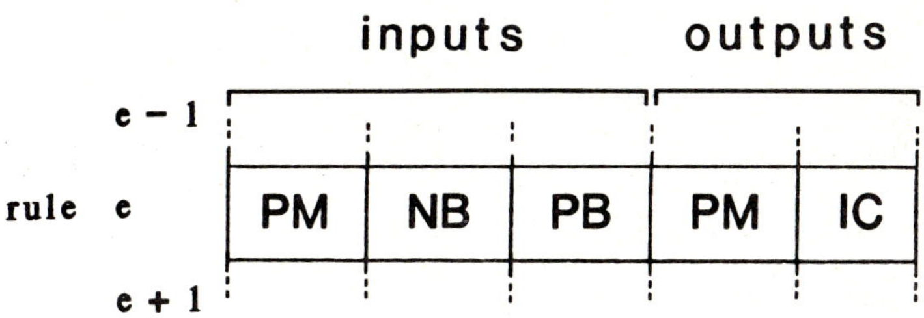

Fig. 4. Tabular Representation of Rule e in Decision Table.

implemented as one-dimensional arrays. Hence, each cell in the table is an array. Although the rule set is structured as a single table, it, in effect, consists of two separate tables: one containing the rules for the computation of the engine rpm adjustment and the other for the elevator adjustment. Computationally, the decision table is treated as two separate tables; the two outputs are computed independently.

The inputs to the controller (rate of descent, airspeed, glide slope) can be considered to be state variables that specify the state of the system, i.e., the airplane, at a given time (time of sampling). To determine what control actions must be taken, the controller logic must first determine which decision rules are applicable given the present state of the system. This rule-evaluation operation is performed using max-min composition.

Given that X and Y are fuzzy subsets of U and V, respectively (i.e., $\mu_X (u_i) : U \Rightarrow [0,1]$, $\mu_Y (v_j) : V \Rightarrow [0,1]$), the decision rule

IF input = X THEN output = Y (6)

constitutes a fuzzy relation

$$R = X \times Y \tag{7}$$

from which it is possible to infer Y = Y' (where $\mu_{Y'} (v_j) : V \Rightarrow [0,1]$ given the input X = X' (where $\mu_{X'} (u_i) : U \Rightarrow [0,1]$). This result is achieved through the compositional operation

$$Y' = X' \circ R. \tag{8}$$

Substitution of Eq. (7) into (8) yields

$$Y' = X' \circ (X \times Y). \tag{9}$$

Since

$$\mu_{X \times Y}(u_i, v_j) \triangleq \mu_X(u_i) \wedge \mu_Y(v_j) \tag{10}$$

and

$$\mu_{X \circ Y}(v_j) \triangleq \bigvee_{i=1}^{m} \mu_X(u_i) \wedge \mu_Y(v_j) \tag{11}$$

where m = | U |, it follows that

$$\mu_{Y'}(v_j) = \bigvee_{i=1}^{m} \left(\mu_{X'}(u_i) \wedge \left(\mu_X(u_i) \wedge \mu_Y(v_j) \right) \right). \tag{12}$$

Consider the case where instead of a single input (antecedent) X in the decision rule, there are now p inputs, $X_1, X_2, \ldots, X_p$ (which correspond to multiple inputs to the controller), such that

IF input1 = X_1,
input2 = X_2,
.
.
. (13)
inputp = X_p

THEN output = Y.

It can be shown that the corresponding fuzzy relation is

$$R = (X_1 \times X_2 \times \ldots \times X_p) \times Y \tag{14}$$

and Eq. (9) thus becomes

$$Y' = X' \circ (X_1 \times X_2 \times \ldots \times X_p \times Y). \tag{15}$$

Correspondingly, Eq. (12) becomes

$$\mu_{Y'}(v_j) = \bigvee_{i=1}^{m} \left(\mu_{X'}(u_i) \wedge \left(\mu_{X_1}(u_i) \wedge \mu_{X_2}(u_i) \wedge \ldots \wedge \mu_{Xp}(u_i) \wedge \mu_Y(v_j) \right) \right). \tag{16}$$

Furthermore, since

$$Y' = \bigcup_{j=1}^{n} \mu_{Y'}(v_j) \tag{17}$$

therefore

$$Y' = \bigcup_{j=1}^{n} \bigvee_{i=1}^{m} \left(\mu_{X'}(u_i) \wedge \left(\mu_{X_1}(u_i) \wedge \mu_{X_2}(u_i) \wedge \cdots \wedge \mu_{X_p}(u_i) \wedge \mu_Y(v_j) \right) \right) \quad . \tag{18}$$

Eq. (18) represents the algorithm for the calculation of the compositional rule of inference for a single rule. For multiple rules, every rule must be evaluated. Hence, given k rules

$$Y' = \bigcup_{j=1}^{n} \bigvee_{i=1}^{m} \bigvee_{e=1}^{k} \left(\mu_{X'}(u_i) \wedge \left(\mu_{X_{e,1}}(u_{1,i}) \wedge \mu_{X_{e,2}}(u_{2,q}) \wedge \cdots \wedge \mu_{X_{e,p}}(u_{p,r}) \wedge \mu_{Y_e}(v_j) \right) \right) \quad . \tag{19}$$

Rearrangement of terms yields

$$Y' = \bigvee_{e=1}^{k} \left(\bigvee_{i=1}^{m} \left(\mu_{X'}(u_i) \wedge \mu_{X_{e,1}}(u_{1,i}) \wedge \mu_{X_{e,2}}(u_{2,q}) \wedge \cdots \wedge \mu_{X_{e,p}}(u_{p,r}) \right) \wedge \left(\bigcup_{j=1}^{n} \mu_{Y_e}(v_j) \right) \right) . \tag{20}$$

Let

$$a_e = \bigvee_{i=1}^{m} \left(\mu_{X_1}(u_i) \wedge \mu_{X_{e,1}}(u_{1,i}) \wedge \mu_{X_{e,2}}(u_{2,q}) \wedge \cdots \wedge \mu_{X_{e,p}}(u_{p,r}) \right) \quad . \tag{21}$$

and

$$Y_e = \bigcup_{j=1}^{n} \mu_{Y_e}(v_j) \quad . \tag{22}$$

Substitution of Eqs. (21) and (22) into (20) yields

$$Y' = \bigvee_{e=1}^{k} a_e \wedge Y_e \tag{23}$$

which is the max-min composition algorithm. The output Y' is, in effect, a weighted linear combination of rule output sets, i.e.

$$Y' = a_1 \cdot Y_1 + a_2 \cdot Y_2 + \ldots + a_k \cdot Y_k \tag{24}$$

where "·" and "+" represent "min" and "max" operations, respectively. The coefficient a_e, which has a value between 0 and 1, is a weighting factor that represents the extent to which the given input X' matches the premise X_e of rule e in the decision table. This algorithm is computationally efficient for it only requires pmk + nk elements versus nm^p elements required by the "brute force" approach.

Defuzzification Stage

Two different defuzzification strategies, Center-of-Area (COA) and Mean-of-Maxima (MOM), were utilized and compared.

The COA algorithm is given by

$$V^* = \frac{\sum_{j=1}^{n} v_j \, \mu_{Y'}(v_j)}{\sum_{j=1}^{n} \mu_{Y'}(v_j)} \,. \tag{25}$$

The value V* generated by Eq. (25) represents the center-of-gravity of the membership function given by Y'.

The MOM algorithm, on the other hand, generates a value which corresponds to the maximum grade of membership in Y'. In the case there is more than one maximum value with the same magnitude, the algorithm generates a value that represents the mean of all local maxima in Y'. This algorithm is given by

$$V^* = \sum_{j=1}^{J} \frac{v_j}{J} \tag{26}$$

where

$$\mu_{Y'}(v_j) = \underset{V}{\mathrm{MAX}} \; \mu_{Y'}(v) \tag{27}$$

and

$$J = \left| \{v\} \right| \,. \tag{28}$$

The value V produced by both algorithms is an integer having a value between -10 and +10 (values greater than ±10 are set equal to ±10 — the largest value allowed). This value is then multiplied by a scaling factor that maps the value onto the appropriate magnitude range corresponding to the variable in question. In the case of engine speed adjustment, for example, V* is multiplied by 10. Thus a V* of -7 corresponds to an engine speed reduction of -7 x 10 or -70 rpm. As will be shown later, the two defuzzification algorithms yield different output results.

Flight Simulator

A computer model, implemented in ALGOL, was developed to simulate the flight of a single-engine (650 bhp @ 2400 rpm), propeller-driven engine with a gross weight of 8000 lbs and wing area of 300 ft^2. The simulation occurs in simulated time (vs. real time). The simulator approximates the behavior of an aircraft during flight with reasonable realism.

The performance of the aircraft, i.e. speed, altitude, rate of descent, etc., is modelled using a combination of graphic and analytical techniques presented in Ref. [1]. The general procedure and basic set of equations used to compute the aircraft performance are as follows:

Step 1. Given current speed n, compute engine brake power P_{br} :

$$P_{br} = 6.73 + 227.53n - 0.05n^2 + 0.0000076n^3 \,. \tag{29}$$

Step 2. Compute propeller power coefficient C_p :

$$C_p = \frac{P_{br}}{\rho n^3 d} \tag{30}$$

where ρ is the air density (0.0024 slug/ft^3) and d is the propeller diameter (10.5 ft).

Step 3. Compute propeller advance ratio J:

$$J = 802.40 - 5.90C_p + 0.045C_p^2 - 0.0053C_p^3 . \tag{31}$$

Step 4. Compute propeller efficiency η :

$$\text{If } J \leq 0.648 \qquad \eta = -0.0029 + 2.52J - 1.94J^2 \tag{32}$$

$$\text{else} \qquad \eta = 164.37 - 704.71J + 1011.10J^2 - 483.09J^3 . \tag{33}$$

Step 5. Compute available power P_{av} :

$$P_{av} = \eta P_{br} . \tag{34}$$

Step 6. Given change in elevator angle $\delta\theta_e$ (input to controller), compute change in aircraft pitch angle $\delta\theta_p$:

$$\delta\theta_p = 0.27 + 0.76e^{0.87\delta\theta_e} \sin(2.27\delta\theta_e - 0.37) . \tag{35}$$

Step 7. Compute new aircraft pitch angle $\theta_{p(new)}$:

$$\theta_{p(new)} = \theta_{p(old)} + \delta\theta_p . \tag{36}$$

Step 8. Compute wing angle of attack θ_w:

$$\theta_w = \theta_p + \theta_{v(old)} \tag{37}$$

where $\theta_{v(old)}$ is the aircraft velocity vector angle.

Step 9. Compute current aircraft speed V:

$$V = Jnd. \tag{38}$$

Step 10. Compute required aircraft power P_{re}:

If $\theta_w \leq 16$

$$P_{re} = 282802 - 524800\theta_w + 151500\theta_w^2 - 42640\theta_w^3 + 655.400\theta_w^4 - 44.790\theta_w^5 + 1.090\theta_w^6 \tag{39}$$

else

$$P_{re} = -3196798 + 6485000\theta_w - 373900\theta_w^2 + 90.960\theta_w^3 + 49.570\theta_w^4 - 1.030\theta_w^5 . \tag{40}$$

Step 11. Compute aircraft rate-of-descent $\frac{\delta A}{\delta t}$:

$$\frac{\delta A}{\delta t} = -24.48 + 0.015n - \sqrt{1 + 0.7x^2}$$

where

$$x = 9.97 + 0.13(|n - 37.63|) . \tag{42}$$

Step 12. Compute new angle between aircraft velocity vector and horizontal $\theta_{v(new)}$:

$$\theta_{v(new)} = \arctan \left[\frac{\frac{\delta A}{\delta t}}{V \sqrt{1 - \left(\frac{1}{V} \frac{\delta A}{\delta t}\right)^2}} \right] . \tag{43}$$

Eqs. (29) through (43) are the actual equations used in the computer flight simulation program. A substantial amount of logic had to be added to the program to handle special situations encountered during flight. The program required a significant amount of "fine tuning" to achieve a reasonable simulation of aircraft flight.

The same pilot who was consulted to generate the decision rules was asked to "try out" the flight simulator. This was done to ensure a consistency between the decision rules and the environment in which they were to be applied. That is, the simulator has to approximate the flight conditions the pilot encounters during actual flight in order for the decision rules to be valid. To avoid any unintentional bias in the formulation of the decision rules, the pilot evaluated the simulator only after he had provided the data for the decision rules.

RESULTS AND DISCUSSION

Three test cases were analyzed. All three cases simulated controller operation of the aircraft down to an altitude of about 250 ft., at which altitude the pilot takes over control. The control objectives were an airspeed of 75 mph, a rate of climb of -350 fpm and a glide-slope angle of 3.0 degrees.

Test Case 1 examines controller performance using COA and MOM defuzzification with 125 decision rules. The airspeed response of both the COA and MOM controllers (Fig. 5a) is overdamped. With the COA controller, the airspeed converges to within six percent of the objective value. In actual flight, this speed error is inconsequential (steady-state errors such as this can be reduced by incorporating an integral term in the control algorithm). The rate-of-climb response of both controllers (Fig. 5b) is also overdamped but has a negligible steady-state error. The underdamped glide-slope angle response of the controllers (Fig. 5c), with an overshoot of approximately three percent, far exceeds the accuracy required under actual flight conditions. A glide-slope angle error of 0.1 degree is insignificant. The settling times are approximately half the duration of the entire landing approach. Although seemingly long, these settling times are quite acceptable in actual conditions. The aircraft pitch angle (Fig. 5d) initially increases and eventually stabilizes at about 7.6 degrees (COA) and 6.8 degrees (MOM). The aircraft descent (Fig. 5e) is very uniform with the COA controller and reasonably uniform with the MOM controller. The runway approach (Fig. 5f) in both cases is extremely smooth. In both cases, a slight oscillation occurs in the last twenty seconds before the controller disengages. This perturbation appears to be due to the fact that an equivalent change in altitude has an increasingly greater effect on the glide-slope angle error as the distance to the runway decreases (e.g. at a distance of 7 mi, a 30 ft drop in altitude changes the glide-slope angle by 0.054 deg; at 1 mi, the same decrease causes a change of 0.32 deg). This situation will lead to overcompensation and, hence, some oscillation. The output of the COA controller is characterized by many incremental adjustments while the MOM controller makes fewer adjustments of larger magnitude. These fluctuations generated by the MOM controller are too severe for actual flight. The fluctuations produced by the COA controller, on the other hand, are acceptable. The performance of both controllers is quite good. The COA controller, however, yields superior results. The COA controller is also desirable because it yields the smaller value of control effort.

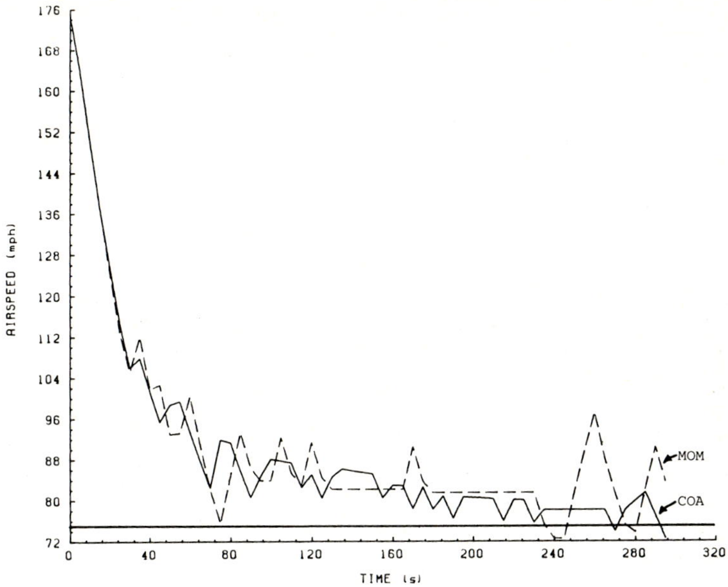

(a). Airspeed.

Fig. 5 Test Case 1 Results: COA vs. MOM Defuzzification.

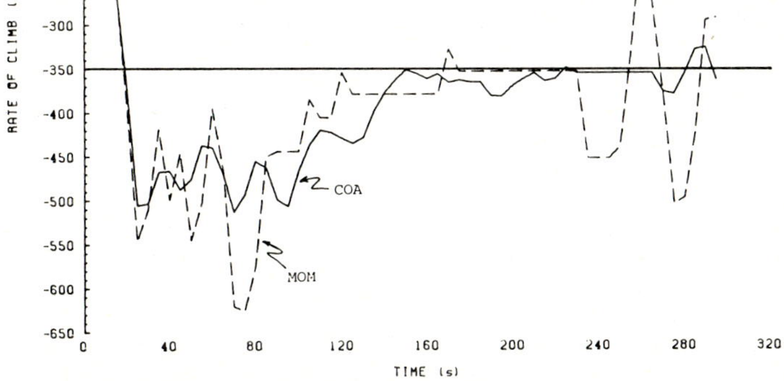

(b). Rate of Climb.

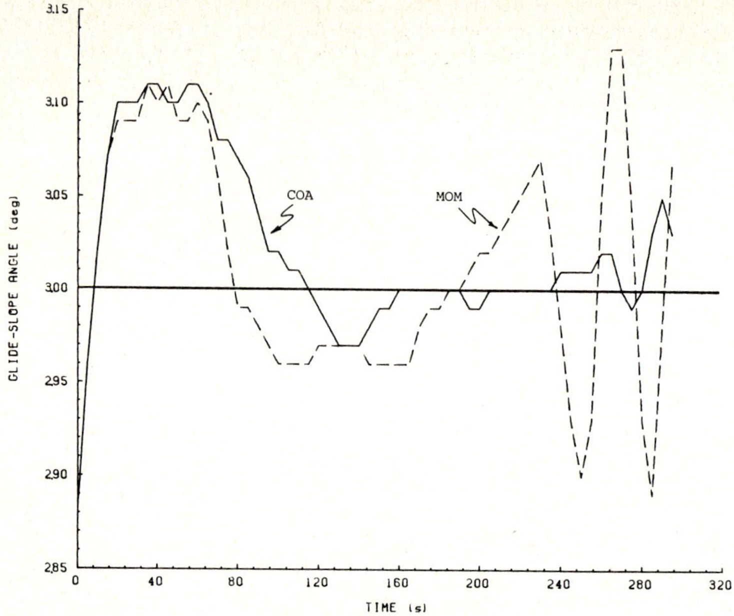

(c). Glide-Slope Angle.

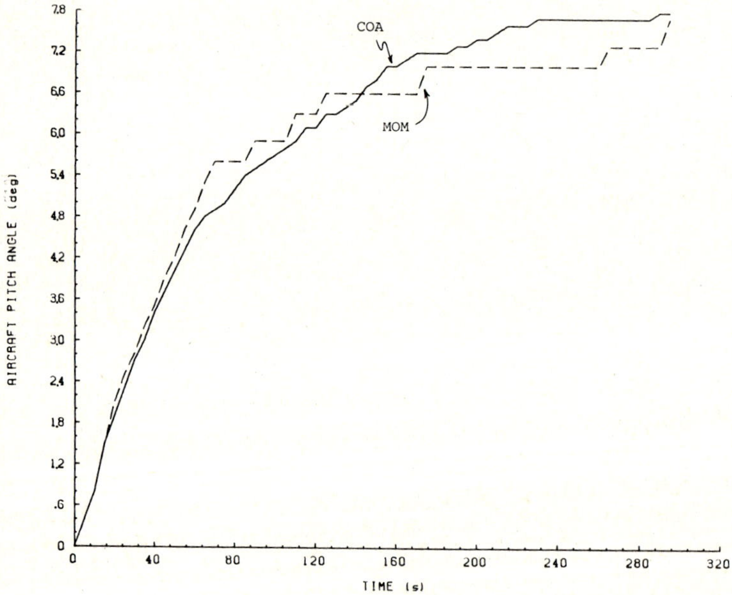

(d). Aircraft Pitch Angle.

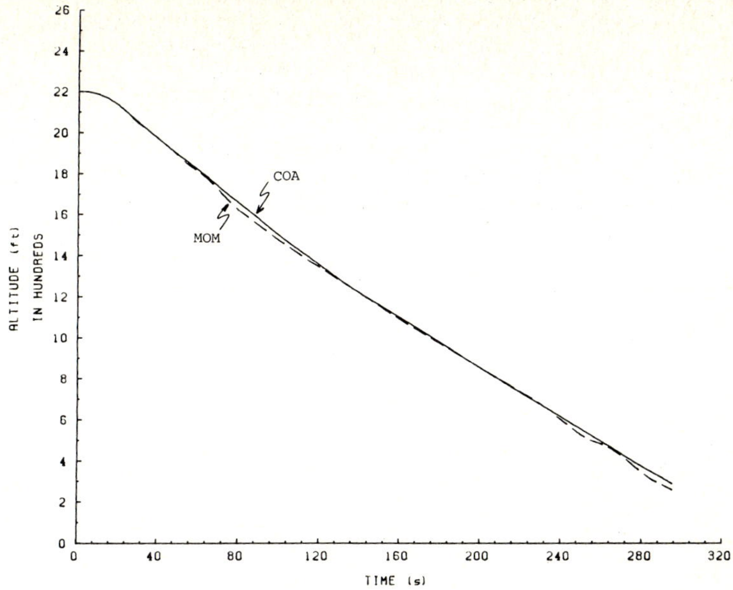

(e). Altitude.

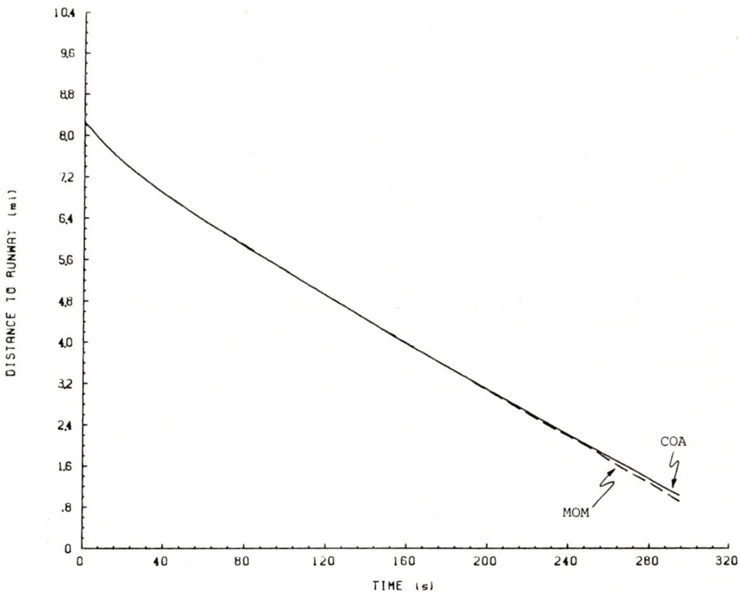

(f). Distance to Runway.

Test Case 2 evaluates the performance of the COA controller based on the quantity of information. The behavior of the controller is examined as the number of decision rules is decreased from 125 to 62, 31 and 15. The rules were reduced by indiscriminately deleting every other rule in the decision table. The results for airspeed, rate of climb and glide-slope angle are shown in Figs. 6a, b and c, respectively. It is interesting to note that the performance of the 62-rule controller is comparable to that of the 125-rule system. Although the steady-state error of the 62-rule controller is larger, it has a smoother response. The performance of the 31-rule controller is marginal yet stable. The 15-rule controller is unstable — it lacks sufficient information. It is possible that a satisfactory performance may have been obtained with fewer than 62 rules had the rules that were retained been chosen more judiciously.

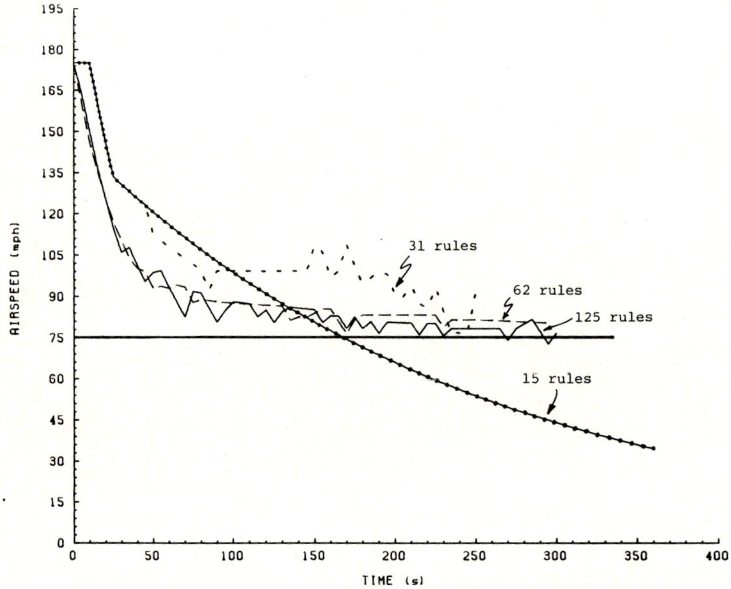

(a). Airspeed.

Fig. 6. Test Case 2 Results: 125 rules vs. 62 rules vs. 31 rules vs. 15 rules.

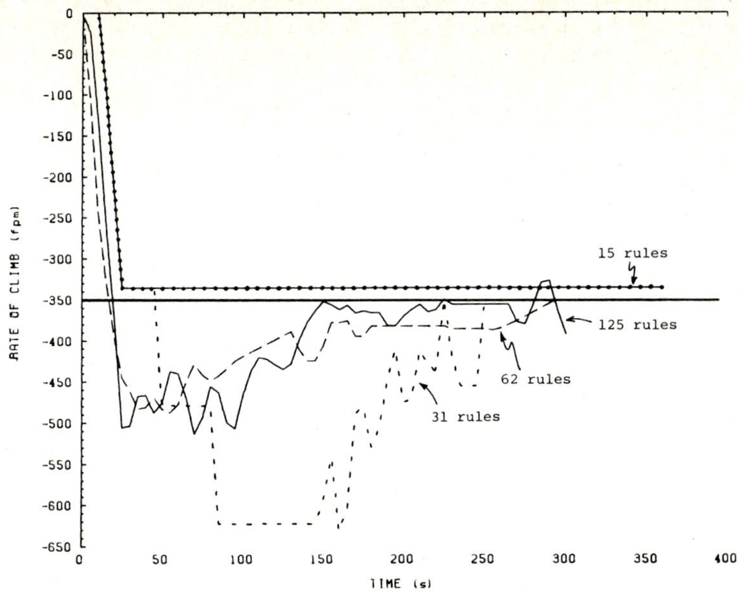

(b). Rate of Climb.

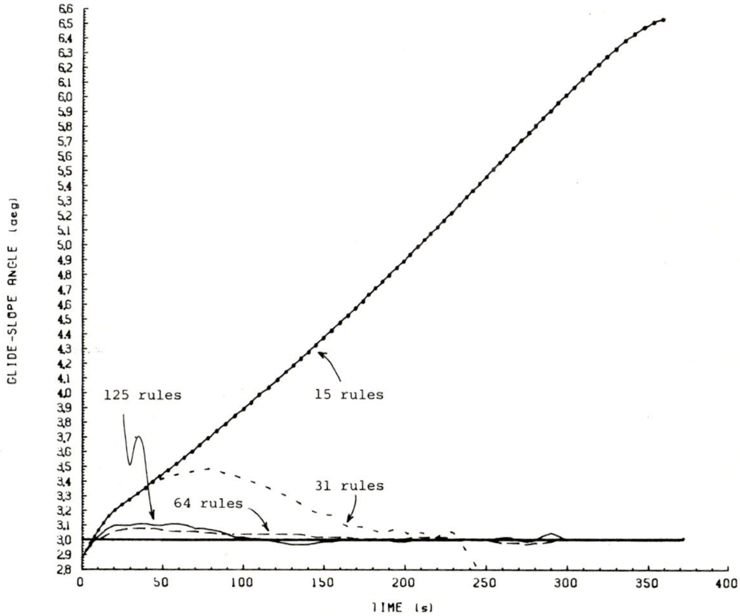

(c). Glide-Slope Angle.

Test Case 3 examines the performance of the COA controller based on the quality of information. Three different degrees of fuzziness were used for all fuzzy variables: high fuzziness (1/-4 + 1/-3 + 1/-2 + 1/-1 + 1/0 + 1/1 + 1/2 + 1/3 + 1/4), medium fuzziness (0.08/-4 + 0.32/-3 + 0.68/-2 + 0.92/-1 + 1/0 + 0.92/1 + 0.68/2 + 0.32/3 + 0.08/4) and low fuzziness (0.5/-1 + 1/0 + 0.5/1). Results are shown in Figs. 7a, b and c.

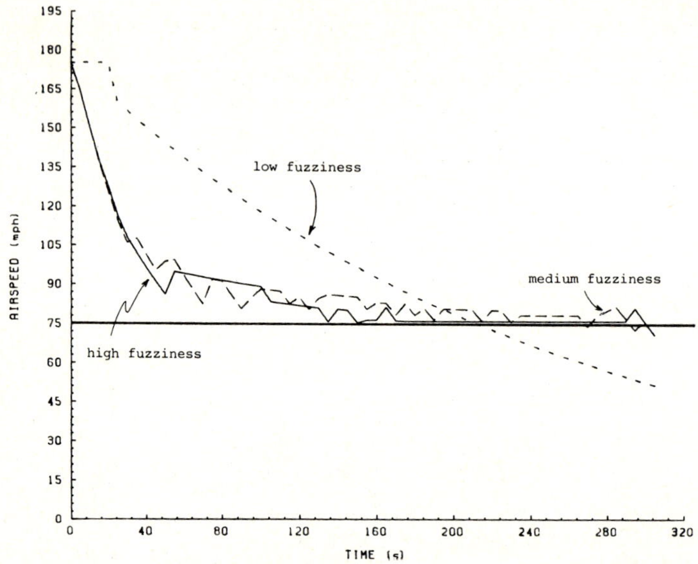

(a). Airspeed.

Fig. 7. Test Case 3 Results: High vs. Medium vs. Low Fuzziness.

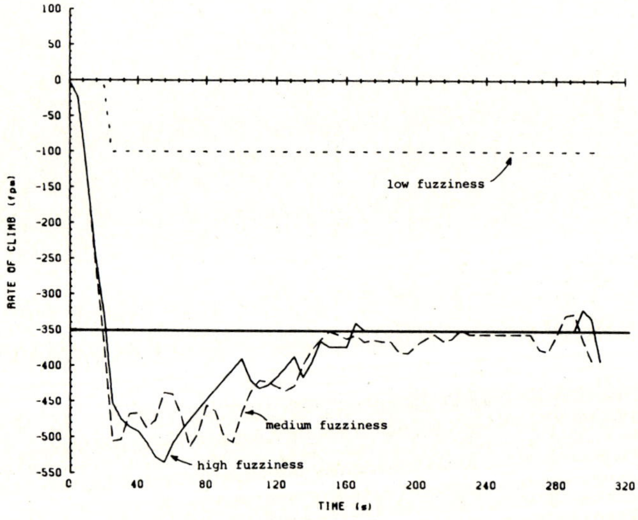

(b). Rate of Climb.

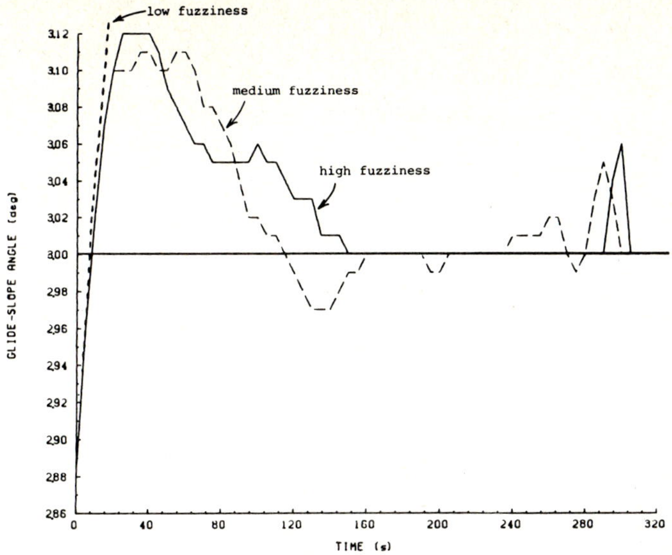

(c). Glide-Slope Angle.

ACKNOWLEDGMENTS

The author wishes to express his sincere appreciation to Dr. Piero Bonissone (General Electric Co.) for his invaluable guidance and to Mr. Robert Schein (General Electric Co.) for his enthusiastic assistance in the development of the flight simulator and decision rules. The author is particularly grateful to Dr. Lotfi Zadeh (UC, Berkeley) for his continued encouragement of this research.

REFERENCES

1. von Mises, R., **THEORY OF FLIGHT** (Dover Publications, New York, 1945).

2. Cauter, G., **INTERNATIONAL AIRPORT** (Octopus Books, London, 1980).

3. Elgerd, O.I., **CONTROL SYSTEMS THEORY** (McGraw-Hill, New York, 1967).

4. Franklin, G.F. and Powell, J.D., **DIGITAL CONTROL OF DYNAMIC SYSTEMS** (Addison-Wesley, New York, 1980).

5. Bonissone, P.P., Cartesian Product and Max-Min Composition, Lecture Notes for Fuzzy Sets in Computer Engineering, Rensselaer Polytechnic Institute (1982).

6. Zadeh, L., Outline of a New Approach to the Analysis of Complex Systems and Decision Processes, IEEE Transactions on Systems, Man, and Cybernetics, SMC-3, No. 1 (1973) 28-44.

7. King, P.J. & Mandami, E.H., The Application of Fuzzy Control Systems to Industrial Control, AUTOMATICA 13 (1977) 235-242.

8. Tong, R.M., A Control Engineering Review of Fuzzy Systems, AUTOMATICA 13 (1977) 559-569.

9. Tong, R.M., Fuzzy Control Systems: A Retrospective. Proceedings ACC-83 (San Francisco, July 1983).

Industrial Applications of Fuzzy Control
M. Sugeno (ed.)

AUTOMOBILE SPEED CONTROL SYSTEM USING A FUZZY LOGIC CONTROLLER

Shuta Murakami and Mikio Maeda

Department of Computer Science
Kyushu Institute of Technology
1-1 Sensuicho, Tobata, Kitakyushu 804
Japan

This paper deals with the synthesis of a fuzzy logic controller by using fuzzy reasoning based on Lukasiewicz logic as proposed by Y. Tsukamoto and its application to automobile speed control system. The controller is composed of a set of linguistic control rules which are conditional linguistic statements expressing the relationship between inputs and outputs.
Two algorithms to compute the values of the manipulated variable on the microprocessor are developed in order to make a real time speed control of automobile. We discuss the results of computer simulations and road tests of an automobile equipped with the microprocessor and we also discuss the usefulness of fuzzy logic controller applied to the automobile speed control system.

INTRODUCTION

Since a fuzzy logic control was proposed by E. H. Mamdani (1974), many studies applied to industrial process such as heat exchanger process, warm water plant, sinter plant and so on have been done. To synthesize a fuzzy controller, so called linguistic control rules are used. These rules are composed by implications such as "if X (output) is A then y (control) is B", which are described by fuzzy relations, $R = A \times B$. Given implicatins and a linguistic proposition, that is, "x is A'", then "y is B'" is inferred by $B' = A' \circ R$. This fuzzy reasoning is called as compositional rule of inference(Zadeh (1973)).

In this paper, we synthesize a fuzzy logic controller by using the fuzzy reasoning method based on linguistic truth value (LTV) proposed by Y. Tsukamoto (1979). We apply it to an automobile speed control system. The fuzzy controller is composed of a set of linguistic control rules which are conditional linguistic statements of relationship between inputs and outputs. The inputs are the 'error' which is the difference between the present speed of automobile and the set point, that is, the desired speed, the first difference of 'error' and the second difference of 'error'. The output, that is the manipulated variable, is inferred by using truth qualification, converse of truth qualification and fuzzy modus ponens.

Two algorithms to calculate the values of manipulated variable on the microprocessor (Intel 8085) are developed. One uses the 'error' and the first difference of 'error' as the input information, the other uses the 'error', the first difference and the second difference of 'error'. These algorithms are implemented in machine language in order to make a real time speed control of automobile. We carried out the road tests by an automobile (TOYOPET CROWN '72, automatic) equipped with the microprocessor and recorders. The characteristics of each fuzzy logic controller are investigated through these experiments and computer simulations. Finally, we discuss the usefulness of fuzzy logic controller developed here which is applied to the automobile speed control system.

FUZZY REASONING [3]

Linguistic Truth Value

Let a fuzzy proposition, P, which includes fuzzy predicates, be described as follows,

$$P = \text{"X is A"},$$

where X is the name of an object and A represents an attributional property of X and A is given as a fuzzy subset of universe of discourse,U. Now, the truth value of fuzzy proposition P is denoted by $\underline{P} = \tau$, which is defined as a fuzzy subset of a universe of truth values, $V \in [0,1]$. If τ is represented in a linguistic way, it is called the linguistic truth value. Examples of linguistic truth values are true, false, very true, more or less true, rather false and so on.

Truth Qualification (TQ) and Converse of Truth Qualification (CTQ)

Given a linguistic proposition of the form

$$P \triangleq \text{"X is A" is } \tau ,$$

TQ is used to obtain the fuzzy predicate, B, equivalent to the following proposition,

"X is B",

where A and B are fuzzy subsets of U. Let μ_A and μ_B be membership functions mapping form U to V. Then, using the diagram of Truth Qualification as shown in Fig.1, $\mu_B(u)$ is given as follows,

$$\mu_B(u) = \mu_\tau(\mu_A(u)) \qquad u \in U$$

Conversely, it is called as Converse of Truth Qualification to find a linguistic truth value provided that A and B are specified. In this case, using the diagram of Converse of Truth Qualification shown in Fig.2, $\mu_\tau(v)$ is given as follows,

$$\mu_\tau(v) = \mu_B(\mu_A^{-1}(v))$$

where μ_A is one-to-one mapping form U to V.

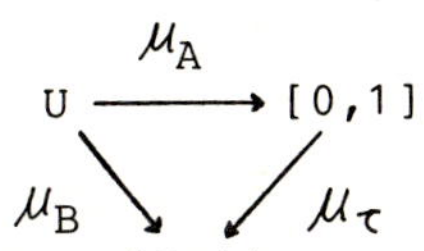

Fig.1 Diagram of TQ

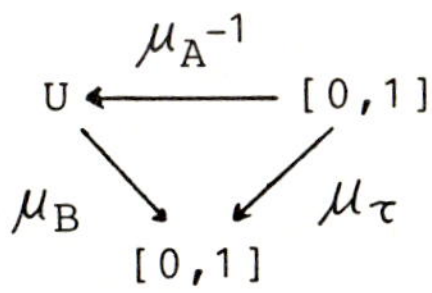

Fig.2 Diagram of CTQ

Fuzzy Reasoning (Fuzzy Modus Ponens)

Let the numerical truth values of proposition, P, Q and R ($\triangleq P \to Q$) be denoted as $|P|$, $|Q|$ and $|R|$, respectively. By using the Lukasiewicz infinite valued logic, the numerical truth value of implication, R, is given as

$$| P \to Q | = (1 - |P| + |Q|) \wedge 1$$

when the truth status of propositions, P and Q, are given by linguistic truth values, the following representation is obtained,

$$\mu_{\underline{P \to Q}}(r) = \sup_{(p,q) \in f^{-1}(r)} [\mu_{\underline{P}}(p) \wedge \mu_{\underline{Q}}(q)]$$

where $f(p,q) \triangleq (1 - p + q) \wedge 1$ and $p, q, r \in [0,1]$.

Fuzzy modus ponens is used to infer the truth status of Q, $\underline{Q}$, given the truth status of implication R, $\underline{R}$, and proposition P, $\underline{P}$. Let the α-level set of $\underline{P}$ and $\underline{R}$ be as follows,

$$\underline{P}^{\alpha} = [p_1(\alpha), p_2(\alpha)], \qquad p_1(\alpha), p_2(\alpha), \alpha \in [0,1]$$

$$\underline{R}^{\alpha} = [r_1(\alpha), 1], \qquad r_1(\alpha), \alpha \in [0,1]$$

then the α-level set of $\underline{Q}$ is obtained as

$$\underline{Q}^{\alpha} = [(p_1(\alpha) - (1 - r_1(\alpha)) \vee 0, 1] \qquad (1)$$

where R is assumed to be possible true and convex, and also P is assumed to be normal and convex. $\mu_{\underline{Q}}(q)$ is given form $\underline{Q}^{\alpha}$ as follows,

$$\mu_{\underline{Q}}(q) = \sup_{\alpha \in [0,1]} [\alpha \wedge \chi_{\underline{Q}^{\alpha}}(q)] \qquad (2)$$

where $\chi_{\underline{Q}^{\alpha}}(q)$ is the characteristic function defined as

$$\chi_{\underline{Q}^{\alpha}}(q) = \begin{cases} 1, & \text{if } \mu_{\underline{Q}}(q) \geq \alpha \\ 0, & \text{if } \mu_{\underline{Q}}(q) < \alpha \end{cases}$$

PROCEDURE TO OBTAIN FUZZY CONTROL

Assume that the linguistic control rules are given by the following forms,

$$R_i \triangleq \text{If } x \text{ is } P_i \text{ then } y \text{ is } Q_i$$

R_i is $\tau_i \Leftrightarrow \underline{R}_i = \tau_i \quad (i = 1,2,..,n)$

where x is the input into the fuzzy controller and y is output from the fuzzy controller. P_i and Q_i are fuzzy subsets on the supporting sets, X and Y, respectively. Now, when the input into the fuzzy controller is given as follows.

"x is D",

let's describe the procedure to obtain the output of fuzzy controller,

1) Let "x is D" be equivalent to "x is P_i" is $\underline{P}_i$. Now that the membership function of D and P_i are specified, by using CTQ, the linguistic truth value of premise, that is, $\underline{P}_i$, is given as follows,

$$\underline{P}_i = \int_V \mu_D(\mu_{P_i}^{-1}(v))/v$$

2) By using fuzzy modus ponens (Eq.(1) and (2)), the linguistic truth value of conclusion, $\underline{Q}_i$, is determined, since $\underline{P}_i$ and $\underline{R}_i$ have been obtained.
3) The fuzzy set, C_i, on the output (control) supporting set, Y, is obtained by using TQ. Now that we have

"y is C_i" = "y is Q_i" is $\underline{Q}_i$,

we get

$$C_i = \int_Y \mu_{\underline{Q}_i}(\mu_{Q_i}(y))/y$$

4) Operating D to all R_i (i = 1,2,..,n), output (control) is determined by the intersection of the fuzzy sets C_i (i = 1,2,..,n) as

$$C = \bigcap_i C_i$$

5) If a non-fuzzy value of output is required, then we may choose y^* for which

$$\mu_C(y^*) = \operatorname*{Max}_{y \in Y} \mu_C(y)$$

This is the basic procedure to determine an actual control by using fuzzy informations.

ALGORITHM TO DETERMINE THE MANIPULATED VARIABLE OF THE AUTOMOBILE SPEED CONTROL SYSTEM

First, we mention the outline of the automobile speed control system which is shown in Fig.3. The speed of the vehicle is detected by the tachogenerator and is converted into digital signals through the A/D converter. The digital data of vehicle speed is taken into the fuzzy controller. The fuzzy controller hardware consists of CPU (Intel 8085), 4K ROM and 16K RAM. This microprocessor calculates the manipulated variable according to the procedure mentioned in the previous section. The calculated value is then converted into an analogue signal through D/A converter. By this analogue signal, the actuator is manipulated and the throttle value of the vehicle is opened or closed. Hence, the speed of the vehicle is controlled.

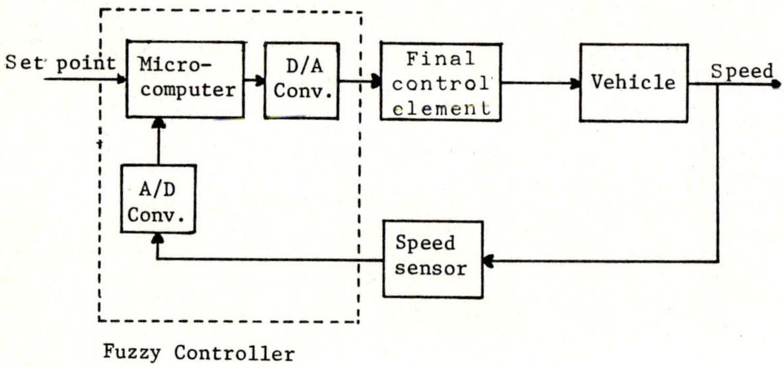

Fig.3 Automoble Speed Control System
Using Fuzzy Logic Controller

Fig.4 shows the block diagram of the final control element which is used in road tests of the automobile speed control system. In the following sections, we will show two algorithms to realize the fuzzy logic controller.

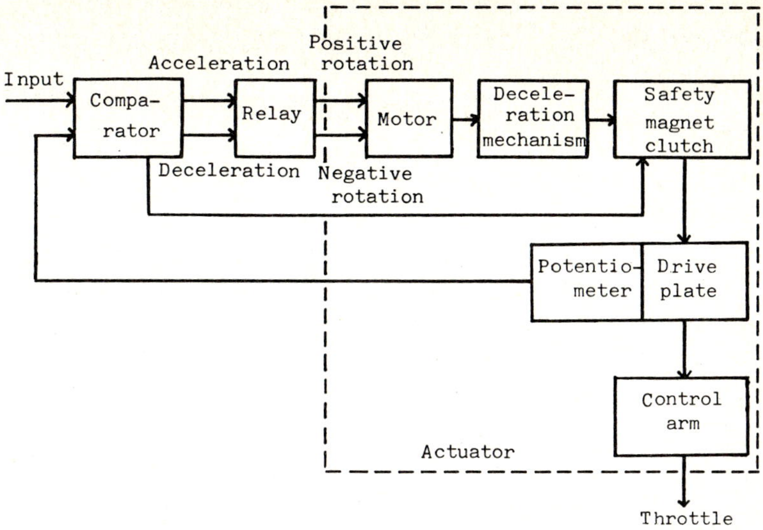

Fig.4 Final Control Element

(A) Case I

We present an algorithm to determine the manipulated variable of the automobile speed control system, when the inputs to the fuzzy controller are the 'error' and the first difference of 'error', where the sampling period is set on 1 second (Murakami (1983)). According to the procedure in section 3, a change in manipulated variable is inferred. Hence, the manipulated variable at the present instant is obtained by adding the one at the last instant to the change in manipulated variable.

The following linguistic control rules of fuzzy logic controller are used in this case

LCR-1 : If e_k is P_e, then Δu_k is PU_e

LCR-2 : If e_k is N_e, then Δu_k is NU_e

LCR-3 : If Δe_k is $P_{\Delta e}$, then Δu_k is $PU_{\Delta e}$

LCR-4 : If Δe_k is $N_{\Delta e}$, then Δu_k is $NU_{\Delta e}$

where

$$e_k = r - x_k,$$

$$\Delta e_k = e_k - e_{k-1},$$

$$\Delta u_k = u_k - u_{k-1},$$

and r is the desired speed (set point), x_k is the speed at time k, and u_k is the manipulated variable at time k. And P., N., PU., NU. are fuzzy subsets on the corresponding supporting sets. Letters, P and PU, imply "positive", and N and NU imply "negative". Each linguistic control rule is considered to be derived form respective goal which can be competitive to each other. An actual control will be determined through a compromise among different criteria [3]. Let us suppose that the truthness of each rule is true. Comparing these rules with a PID controller, we can say that LCR-1 and LCR-2 correspond to the integral action and that LCR-3 and LCR-4 correspond to the proportional action.

Now, the membership functions of P., N., PU., and NU. are to be characterized as follows,

$$\mu_{P_e}(e) = \frac{1}{\pi}\tan^{-1}(a\cdot e) + 0.5, \quad \mu_{N_e}(e) = \frac{1}{\pi}\tan^{-1}(-a\cdot e) + 0.5,$$

$$\mu_{P_{\Delta e}}(\Delta e) = \frac{1}{\pi}\tan^{-1}(a'\cdot\Delta e) + 0.5, \quad \mu_{N_{\Delta e}}(\Delta e) = \frac{1}{\pi}\tan^{-1}(-a'\cdot\Delta e) + 0.5,$$

$$\mu_{PU_e}(\Delta u) = \frac{1}{\pi}\tan^{-1}(b\cdot\Delta u) + 0.5, \quad \mu_{NU_e}(\Delta u) = \frac{1}{\pi}\tan^{-1}(-b\cdot\Delta u) + 0.5,$$

$$\mu_{PU_{\Delta e}}(\Delta u) = \frac{1}{\pi}\tan^{-1}(b'\cdot\Delta u) + 0.5, \quad \mu_{NU_{\Delta e}}(\Delta u) = \frac{1}{\pi}\tan^{-1}(-b'\cdot\Delta u) + 0.5,$$

where

$$a = \tan(0.45\cdot\pi)/r,$$

$$a' = \tan(0.45\cdot\pi)/c_{max} \; (\triangleq a/c), \; c = c_{max}/r,$$

$$b = \tan(0.45\cdot\pi)/u_{max},$$

$$b' = \tan(0.45\cdot\pi)/d_{max} \ (\triangleq b/d),\ d = d_{max}/u_{max},$$

and, c_{max}, u_{max}, and d_{max} are scaling factors which are parameters to determine the size of respective supporting set. In Fig.5(a) and (b), the membership functions of fuzzy sets, P_e and N_e, and PU_e NU_e are shown as examples.

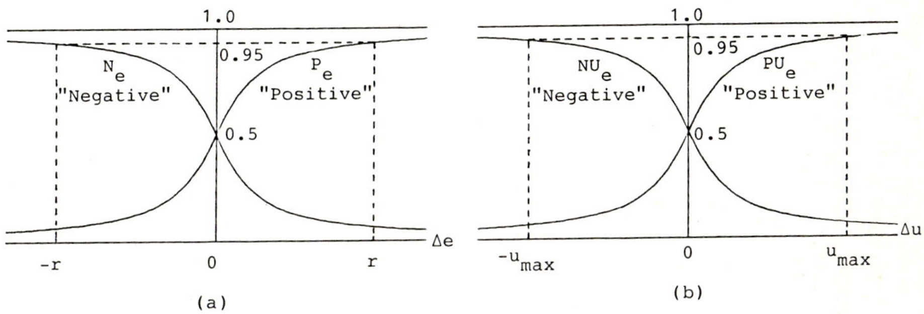

Fig.5 Example of Membership Function in Case I

Given inner data of e_k and Δe_k, we can determine the non-fuzzy control Δu_k^* according to the procedure mentioned in the previous section. To simplify the computation, we assume that the inputs, e_k and Δe_k, are non-fuzzy values. The algorithm to determine the actual control (Δu_k^*) at time k is summarized as follows,

1) If $e_k>0$ and $\Delta e_k>0$, and

i) $a\cdot e_k = c/(a\cdot\Delta e_k)$, then $\Delta u_k^* = \sqrt{d}/b$

ii) $\lambda>0$, then $\Delta u_k^* = (-(1+d) + \sqrt{(1+d)^2 + 4\lambda^2 d})/2\lambda b$

iii) $\lambda<0$, then $\Delta u_k^* = (-(1+d) - \sqrt{(1+d)^2 + 4\lambda^2 d})/2\lambda b$

iv) $(a/b)\ e_k = (a\ d/b\ c)\cdot\Delta e_k$, then $\Delta u_k^* = (a/b)\cdot e_k$

2) If $e_k<0$ and $\Delta e_k<0$, and

i) $a\cdot e_k = c/(a\cdot\Delta e_k)$, then $\Delta u_k^* = -\sqrt{d}/b$

ii) $\lambda > 0$, then $\Delta u_k^* = (-(1 + d) - \sqrt{(1 + d)^2 + 4\lambda^2 d})/2\lambda b$

iii) $\lambda < 0$, then $\Delta u_k^* = (-(1 + d) + \sqrt{(1 + d)^2 + 4\lambda^2 d})/2\lambda b$

iv) $(a/b)\cdot e_k = (a\, d/b\, c)\cdot \Delta e_k$, then $\Delta u_k^* = (a/b)\cdot e_k$

3) If $e_k \geq 0$ and $\Delta e_k \leq 0$, or $e_k \leq 0$ and $\Delta e_k \geq 0$, and

i) $\lambda = 0$, then $\Delta u_k^* = 0$

ii) $\lambda \neq 0$, then $\Delta u_k^* = (-(1 + d) + \sqrt{(1 + d)^2 + 4\lambda^2 d})/2\lambda d$

where

$$\lambda = (a\cdot e_k + (a/c)\cdot \Delta e_k)/(1 - a\cdot e_k\ (a/c)\cdot \Delta e_k)$$

Using the above algorithm, the manipulated variable at time k, u_k, which is the input into the final control element, is determined as

$$u_k = u_{k-1} + \Delta u_k^*$$

(B) Case II

In this case, as the inputs to fuzzy logic controller, we consider the second difference of 'error' that is, $\Delta^2 e_k$, added in case I. By adding this information, we can expected that the responce time gets faster than in case I. The linguistic control rules used in this case are as follows,

LCR-1 : If e_k is P_e, then Δu_k is PU_e,

LCR-2 : If e_k is N_e, then Δu_k is NU_e,

LCR-3 : If Δe_k is $P_{\Delta e}$, then Δu_k is $PU_{\Delta e}$,

LCR-4 : If Δe_k is $N_{\Delta e}$, then Δu_k is $NU_{\Delta e}$,

LCR-5 : If $\Delta^2 e_k$ is $P_{\Delta^2 e}$, then Δu_k is $PU_{\Delta^2 e}$,

LCR-6 : If $\Delta^2 e_k$ is $N_{\Delta^2 e}$, then Δu_k is $NU_{\Delta^2 e}$, where

$$\Delta^2 e_k = \Delta e_k - \Delta e_{k-1}.$$

The added rules, LCR-5 and LCR-6, correspond to the derivative action in PID controller.

The membership functions of P. and N. for the premise sides are specified as follows.

$$\mu_{P_e}(e) = \frac{1}{\pi}\tan^{-1}(a_1 \cdot e)+0.5, \qquad \mu_{N_e}(e) = \frac{1}{\pi}\tan^{-1}(-a_1 \cdot e)+0.5,$$

$$\mu_{P_{\Delta e}}(\Delta e) = \frac{1}{\pi}\tan^{-1}(a_2 \cdot \Delta e)+0.5, \quad \mu_{N_{\Delta e}}(\Delta e) = \frac{1}{\pi}\tan^{-1}(-a_2 \cdot \Delta e)+0.5,$$

$$\mu_{P_{\Delta^2 e}}(\Delta^2 e) = \frac{1}{\pi}\tan^{-1}(a_3 \cdot \Delta^2 e)+0.5, \quad \mu_{N_{\Delta^2 e}}(\Delta^2 e) = \frac{1}{\pi}\tan^{-1}(-a_3 \cdot \Delta^2 e) +0.5.$$

In Fig.6(a), an example of membership for 'positive' and 'negative' is shown, where $a_i = \tan(0.45 \cdot \pi)/c_i$ $(i = 1,2,3)$.

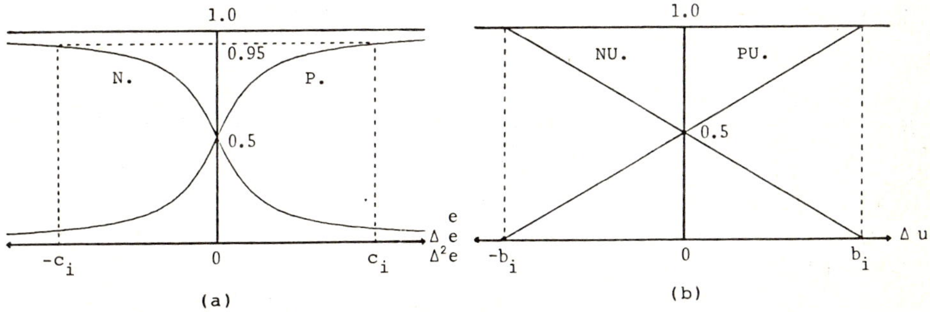

Fig.6 Example of Membership Function in Case II

For the simplicity of calculation, the membership functions of PU. and NU. for the conclusion sides are characterized by linear functions :

$$\mu_{PU_e}(\Delta u) = \frac{1}{2b_1}\Delta u + 0.5, \qquad \mu_{NU_e}(\Delta u) = -\frac{1}{2b_1}\Delta u + 0.5,$$

$$\mu_{PU_{\Delta e}}(\Delta u) = \frac{1}{2b_2}\Delta u + 0.5, \qquad \mu_{NU_{\Delta e}}(\Delta u) = -\frac{1}{2b_2}\Delta u + 0.5,$$

$$\mu_{PU_{\Delta^2 e}}(\Delta u) = \frac{1}{2b_3}\Delta u + 0.5, \quad \mu_{NU_{\Delta^2 e}}(\Delta u) = -\frac{1}{2b_3}\Delta u + 0.5.$$

An example of membership for the conclusion side is shown in Fig.6(b).

Given the input information of e_k, Δe_k, $\Delta^2 e_k$ at time k, the non-fuzzy control, Δu_k^* at time k, is determined according to the procedure in section 3 as follows,

$$\Delta u_k^* = \text{medium}\{\Delta u_1, \Delta u_2, \Delta u_3\},$$

where

$$\Delta u_1 = \frac{1}{\pi}\{\tan^{-1}(a_1 \cdot e_k) + \tan^{-1}(a_2 \cdot \Delta e_k)\}/(d_1 + d_2),$$

$$\Delta u_2 = \frac{1}{\pi}\{\tan^{-1}(a_2 \cdot \Delta e_k) + \tan^{-1}(a_3 \cdot \Delta^2 e_k)\}/(d_2 + d_3),$$

$$\Delta u_3 = \frac{1}{\pi}\{\tan^{-1}(a_3 \cdot \Delta^2 e_k) + \tan^{-1}(a_1 \cdot e_k)\}/(d_3 + d_1),$$

$$d_i = 1/(2b_i) \quad (i = 1,2,3),$$

and medium implies that a middle value is taken out of a set of three values.

The output Δu_k^* is expressed as a linear combination with two of $\tan^{-1}(a_1 \cdot e_k)$, $\tan^{-1}(a_2 \cdot \Delta e_k)$ and $\tan^{-1}(a_3 \cdot \Delta^2 e_k)$. From viewpoint of e_k, Δe_k and $\Delta^2 e_k$, it is represented as a nonlinear combination of them. Hence, this fuzzy logic controller could be said to be a sort of nonlinear PID controller.

SIMULATION

To investigate the characteristics of the fuzzy logic controller developed in this paper, some computer simulations of the automobile speed control system were carried out. As a controlled system, we used an automobile, TOYOPET CROWN ('72, Automatic). The transfer function of the controlled system including the final control element was approximately identified from the step response as follows,

$$G(s) = \frac{35.78}{(1 + 1.728\ s)(1 + 16.848\ S)} e^{-0.4\ s}$$

(A) Case I

Fig.7 shows the result of computer simulation when the set point is set at 60 Km/h for the first 80 seconds and then it is changed to at 40 Km/h. In this case, as the controlled variable (speed) gets very near the set point, the speed once slows down slightly and then settles up at set the point. PI controller could not realize such a response. The changing pattern of manipulated variable shows that it is close to the human operation of speed control. The larger c_{max} and u_{max} are, the faster the rise time becomes as shown in Fig.8. But, in this case, an overshoot occurred.

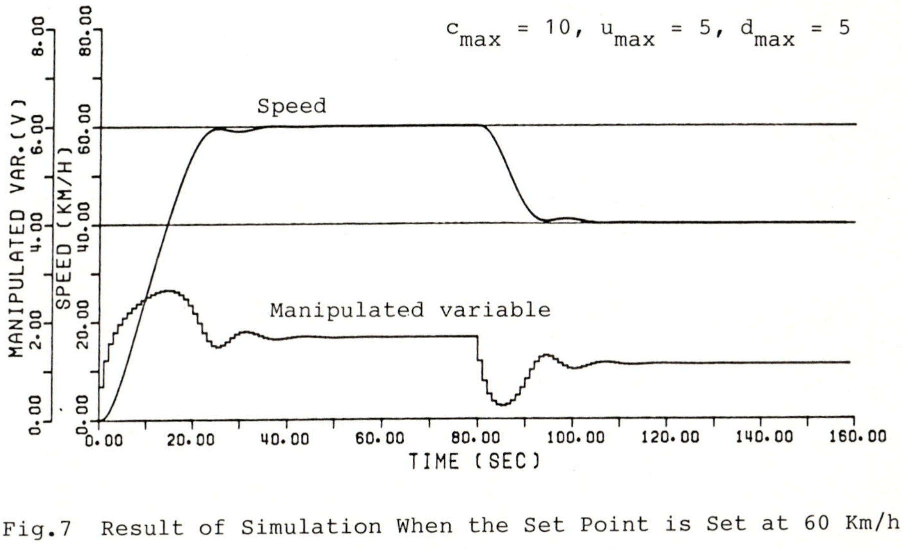

Fig.7 Result of Simulation When the Set Point is Set at 60 Km/h and Then Changed to 40 Km/h in Case I

Fig.9 shows the result of the simulation in the case where the control is suspended between the 81st second and the 95th second and then the control is again started from the 95th second. In this case, very slight overshoot is observed, since the manipulated variable from the 95th second gets larger.

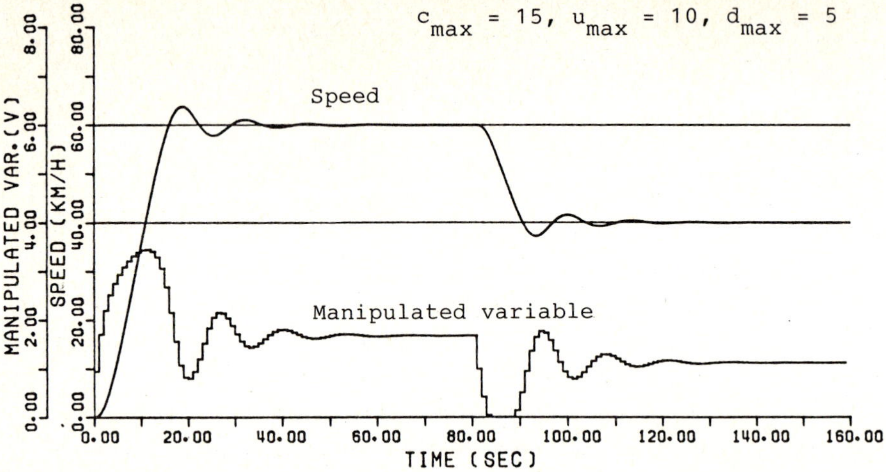

Fig.8 Result of Simulation When Scaling Factors are Changed in Case I

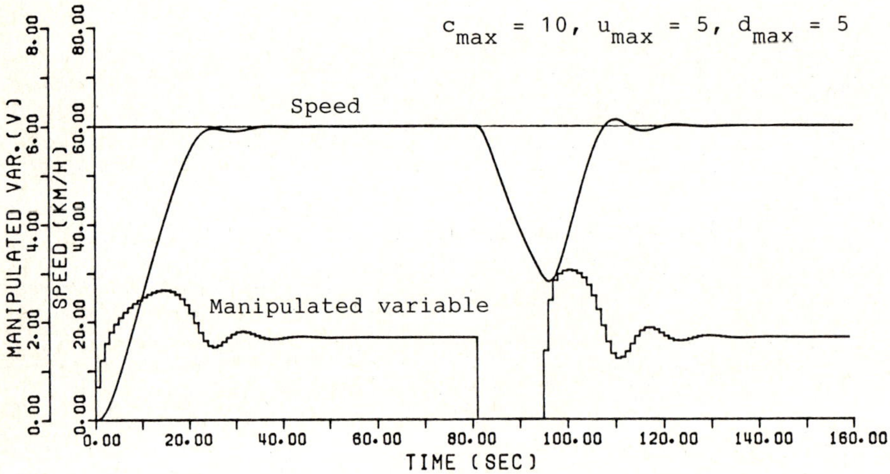

Fig.9 Result of Simulation When Control is Suspended in Case I

Fig.10 shows the result in the case where the gain of the controlled system is forced to change after the 60th second by using a sine curve with amplitude of half the system gain. This corresponds to the case where the vehicle is supposed to run on an uphill and downhill road. In spite of rather big change of gain, the automobile is well controlled by using this fuzzy logic controller.

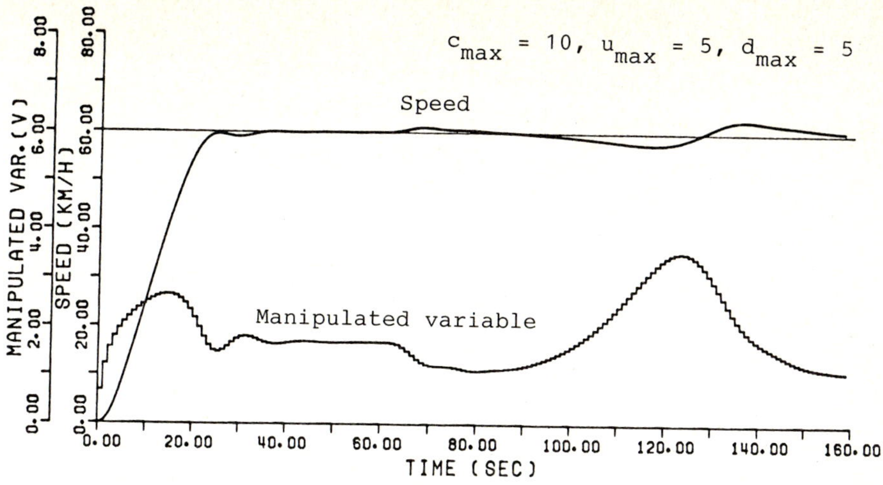

Fig.10 Result of Simulation When System Gain is Forced to Change by Using Sine Curve in Case I

(B) Case II

The fuzzy logic controller in case II is designed to make the response time faster than case I. A computer simulation result is shown in Fig.11, where the heavy line denotes the result for the fuzzy logic controller and thin line for the PID controller. Comparing with the result of Fig.10 in case I, the response time is apparently improved. But, in this case, an overshoot is slightly observed. Comparing the result for the fuzzy controller with that for the PID controller, it can be said that the former controller is superior to the latter in control performance (for example, settling time and rise time), especially in the response for the cyclical change of system gain.

From these simulation results, we could obtain such suggestions that the fuzzy controller is very useful for the automobile speed control system and that a speed control system which is robust to the disturbance and the change of system characteristics can be constructed.

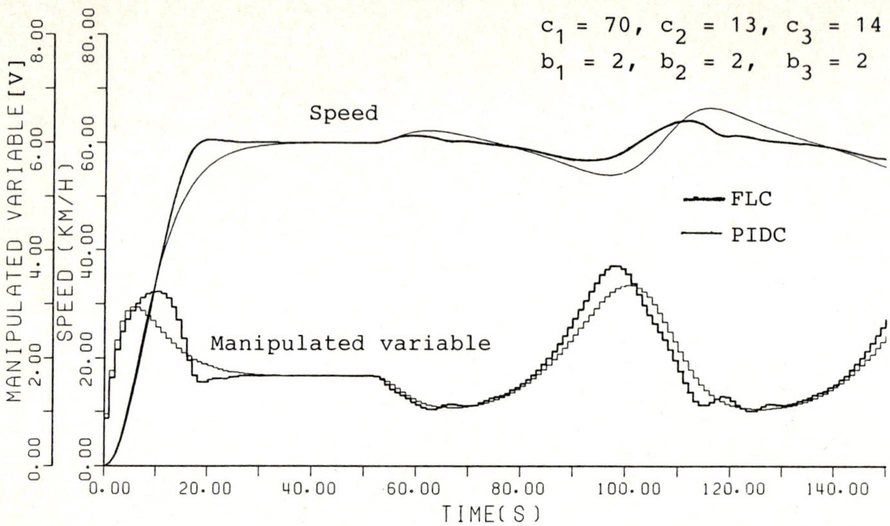

Fig.11 Result of Simulation When System Gain is Forced to Change by Using Sine Curve in Case II

EXPERIMENT

To perform real time road tests, the experimental vehicle was equipped with a microprocessor which consisted of CPU (Intel 8085), 4K ROM and 16K RAM. It was also equipped with devices to record the speed and the manipulated variable. The algorithm to calculate the manipulated variable mentioned in the previous section was programmed using the machine language in ROM. Computing time to obtain a manipulated variable was about 50 msec in both cases.

The first road test was done on the flat road. Fig.12 shows the result in case I. The set point was first set at 40 Km/h and then changed up to 60 Km/h. The response of speed seems to have the same characteristic as the results of the computer simulations. That is, as the speed of vehicle gets very near the set point, it slows down and then settles up to the steady state. And also an overshoot is not observed in both cases. The changes of manipulated variable in the steady state was not too large for the driver to feel on his body.

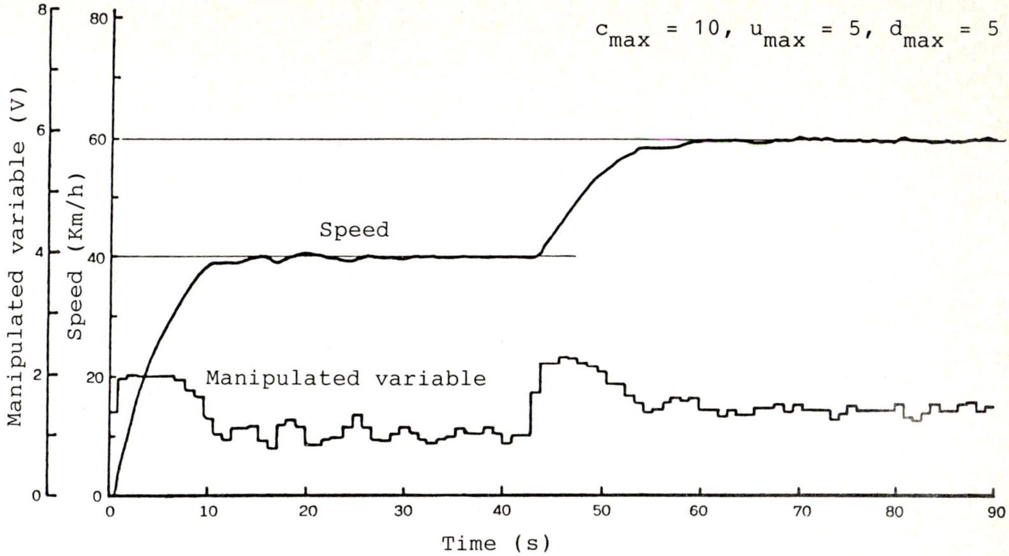

Fig.12 Result of Road Test on Flat Road in Case I

Next, we carried out a road test on the interval from Murasakigawa Interchange to Kurosaki Interchange of Kitakyushu Highway. The distance was about 12 Km. The interval has the uphill and downhill road with the ascent of maximum 4.7% and the descent of maximum 3.7%. In Fig.13 and 14, the result of the road test in case I and case II are shown. The set point was set at 60 Km/h, since the speed limit of the highway is 60 Km/h. The speed of automobile is well controlled even on the uphill and downhill road in both cases. The fluctuation of speed from the set point stays within about $\pm$2Km/h. In the result of case II, a slight overshoot is observed, but the response time is improved ones that in case I.

We find that the value of the manipulated variable in the steady state in the first road test (set point 60 Km/h in Fig.12) and in the third road test in Fig.14 is less than that in the second road test in Fig.13. This is because We used the high octane gasoline in the first and third road test, and the regular gasoline in the second road test by mistake. In spite of the kind of gasoline used, the control performance has not so great difference. From these road test of an automobile speed control system with a fuzzy logic controller, we can conclude that the fuzzy logic controller developed in this paper is robust to the change of gain of controlled vehicle and the disturbances. By using this fuzzy logic controller, a stable and comfortable ride at a desired speed could be realized.

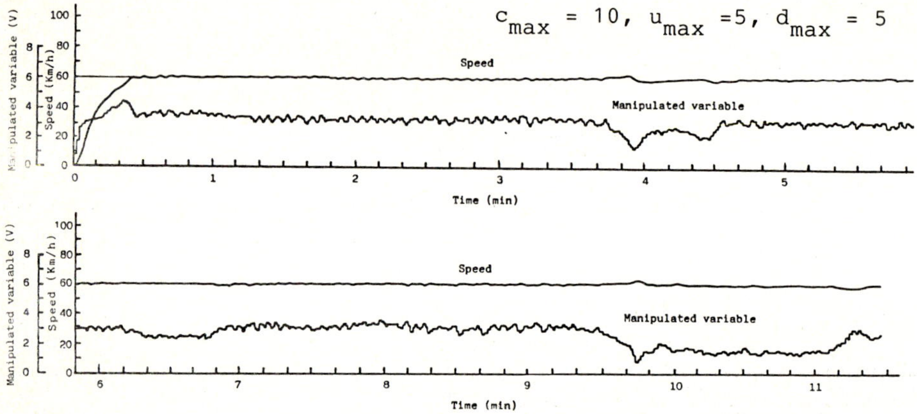

Fig.13 Result of Road Test on Kitakyushu Highway in Case I

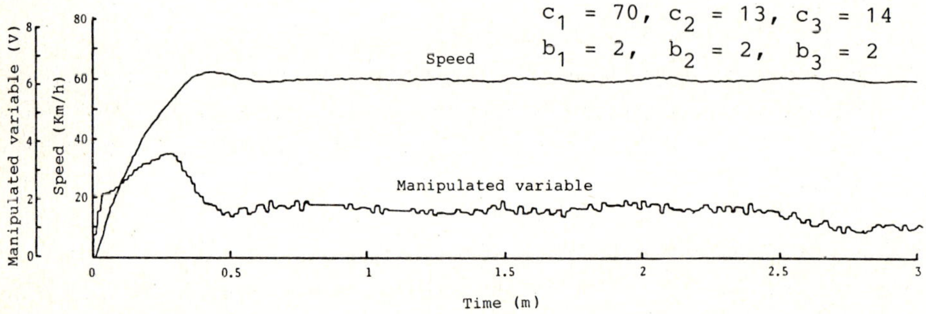

Fig.14 Result of Road Test on Kitakyushu Highway in Case II

CONCLUDING REMARKS

We could synthesize the automobile speed control system using a fuzzy logic controller which has robustness. The results of the road tests show the usefulness of the fuzzy logic controller. The fuzzy controller developed here is said to be suitable for controlling such systems (processes) where the system gain changes frequently and that the characteristics of controlled system are not precisely known.

We are going to develop the fuzzy control system so that it includes the mechanism to learn the driver's driving manner. Also we are to investigate the quantitative properties of scaling factors (parameters to determine sizes of supporting sets) which will affect control performances.

REFERENCES

[1] Mamdani, E. H., Application of Fuzzy Algorithms for Control of a Simple Dynamic Plant, Proc. IEE, 121, 12, (1974), 1585-1588.

[2] Zadeh, L. A., Outline of a New Approach to the Analysis of Complex Systems and Decision Processes, IEEE Trans., Vol. SMC-3, No.1, (1973), 28-44.

[3] Tsukamoto, Y., Fuzzy Logic Based on Lukasiewicz Logic and Its Application to Diagnosis and Control, Doctoral Dissertation (T.I.T.), (1979).

[4] Murakami, S., Application of fuzzy controller to automobile speed control systems, Proc. of the IFAC Symposium on Fuzzy Information Knowledge Representation and Decision Analysis, Marseille (1983) 43-48.

Industrial Applications of Fuzzy Control
M. Sugeno (ed.)

AN EXPERIMENTAL STUDY ON FUZZY PARKING CONTROL USING A MODEL CAR

M. Sugeno and K. Murakami

Department of Systems Science
Tokyo Institute of Technology
4259 Nagatsuta, Midori-ku
Yokohama 227, JAPAN

This paper discusses an application of fuzzy control to parking a car. Fuzzy control rules are derived by modelling a man's parking actions. Experiments are successfully performed using a model car equipped with sensing devices and a micro-processor.

Key words : fuzzy control, car parking, modelling expert's control actions

INTRODUCTION

Since Mamdani's pioneering work [1] on fuzzy control, many studies have been reported in this area. Recently our main interest has been towards the practical applications of fuzzy control to industrial processes [5, 9, 10]. When we apply fuzzy control to an industrial process, one of the key problems to be solved is to find fuzzy control rules, i.e., to design a fuzzy controller. A number of studies have been devoted to this problem [2, 3, 4, 7]. There are, in general, four methods of design. Among them the fourth method, self-organization of a fuzzy controller, is not applicable in a real process control. Let us briefly review the other three methods.

<u>(1) Expert's experience and knowledge</u>

Most fuzzy controllers have been designed so far by referring to a human operator's experience and/or a control engineer's knowledge. In fact we can say that fuzzy contol was the first practical application of the expert system. In many cases where an operator plays an important role in process control, it is very useful to find his know-how on control by interviews and to express it in terms of fuzzy implications. It is also possible for a control engineer to list a number of protocols based on his knowledge about the process to be controlled and also based on his general control engineering sense.

There are, however, some disadvantages in this method of design. Firstly it is often the case that an operator cannot verbalize his knowledge. Secondly it is difficult, even for the control engineer, to write down control rules if the process is too complex. We can also point out that this method is essentially heuristic. So it is very difficult to give a general design procedure.

(2) Modelling operator's control actions

When an operator's skill is important, it is very useful to derive fuzzy control rules by modelling an operator's control actions. For example when we drive a car, we control it directly by using our hands and legs rather than by thinking in our brains. If it is possible to model an operator's control behavior in terms of fuzzy implications using the input-output data concerned with his control actions, we can use the obtained model as a fuzzy controller.

The method of modelling an operator's control actions is basically the same as that of modelling a process. However it is easier to model an operator than to model a process since in the former the input variables of the model are easily found, for example, by asking an operator what kind of informations he uses in his control or by watching his actions. It is also possible to find an approximate number of fuzzy control rules. In a real situation, it is desirable to combine the first and the second methods.

(3) Modelling a process

The first method is somehow based on a rough idea about the characteristics of a process: the output increases as the input does, the process has a time lag, and so on. The second method uses only those variables available to the process operator. We can say that a fuzzy controller designed by the above two methods works well only in the case where an expert plays an important role in process control. This fuzzy controller may work even better than an expert since it does not make error.

If we cannot rely on an operator and if we want to have better results than an operator gives, we have to find another method of design. It is expected that if we have a certain, possibly fuzzy, model of a process, we may design a fuzzy controller in a more sophisticated way. This is of course what control theory is about. Some ideas for deriving fuzzy control rules from a fuzzy model of a process have been studied, provided that a fuzzy model is given [4, 6, 8].

By fuzzy modelling what we mean here is representation of the characteristics of a process by a set of its fuzzy behaviors which are also expressed by using fuzzy implications concerned with inputs, state variables and outputs.

There are two ideas for designing a fuzzy controller based on a fuzzy model. The first one is a heuristic method in which we set a fuzzy control rule to compensate for an undesirable system behavior by considering the control objective. The second one is a method, as in optimal control theory, to determine the structure and the parameters of control rules so that the system with a fuzzy controller satisfies the control objective, for example, minimizes the performance index.

This paper deals with the second idea for designing a fuzzy controller. We apply fuzzy control to the problem of parking a model car in a garage. The model car used in the study is equipped with sensing devices and on board microprocessor as a fuzzy controller. Fuzzy control rules are derived by modelling an expert's way of driving a car.

PRELIMINARIES

Fuzzy implications and reasoning method

Since we are interested in a fuzzy model of the operator's control actions, in this paper we use the following fuzzy implications for a fuzzy controller. The i-th implication is of the form

$$R^i\text{: If } x_1 \text{ is } A_1^i,\ x_2 \text{ is } A_2^i,\ \ldots\ldots,\ x_n \text{ is } A_n^i$$

$$\text{then } y^i = p_0^i + p_1^i x_1 + \ldots\ldots + p_n^i x_n$$

where x_1, x_2, ..., x_n are the input variables, y^i is the output from the i-th implication, A_1^i, A_2^i, ..., A_n^i are fuzzy variables, and p_0^i, p_1^i, ..., p_n^i are the coefficients of a linear equation.

In the sequel the membership function of a fuzzy set is written simply as A(x) which is composed of straight lines. As is seen, the premise of an implication expresses a fuzzy subspace of the input space and the consequence is a linear equational relation which holds in the fuzzy subspace described in the premise. We should note that each implication used in a fuzzy controller may be different from the above general format. Let us consider some examples.

R^1: If x_1 is small, x_2 is small then $y^1 = 2.0 + 1.0x_1 + 3.0x_2 + 0.5x_3$

R^2: If x_1 is big, x_4 is medium then $y^2 = 3.0 + 0.8x_1 + 2.5x_2 + 1.0x_3$

We recognize that some of the input variables of the premises are different from those of the consequents: x_3 does not appear in the premises and x_4 does not appear in the consequents, and so on.

When the inputs x_1^o - x_n^o are given, the truth value of the premise of the i-th rule is calculated as

$$w^i = \prod_{j=1}^{n} A_j^i(x_j^o)$$

and the output y^o is inferred from m rules by taking the weighted average of the y^i's.

$$y^o = \sum_{i=1}^{m} w^i y^i \Big/ \sum_{i=1}^{m} w^i$$

Identification algorithm of fuzzy control rules

Suppose that we are given the input-output data x_{1k}, x_{2k}, ... , x_{nk}, and y_k, $k \in K$ taken from the operator's control actions. We have to identify

(1) the number of fuzzy subspaces in the input space, i.e., the number of fuzzy control rules,

(2) the membership functions of the fuzzy variables in the premises,

(3) the coefficients in the consequences of the rules the number of which is $m \times (n + 1)$.

Those are identified so that the output error is minimized where we mean, by the output error, the difference between the output of a model and the output data. In general there are two kinds of structure identification problems. One is the ordinary problem of finding the necessary variables in the premises and also in the consequents. As far as modelling the operator's control actions is concerned, we can easily find this kind of structure by observing his control situation. So we do not discuss this problem. The other one is concerned with the number of fuzzy subspaces.

Now the first problem, i.e., to find the necessary number of subspaces in the input space, has no general solution since it is a combinatorial problem. However, engineering sense is available and a heuristic search of the optimal fuzzy partition of the input space is also possible.

Provided that (1) the number of fuzzy subspaces and (2) the membership functions are given, (3) the coefficients are easily identified using the output equation of y^o so as to minimize the output error. This problem is nothing but an ordinary parameter identification problem. As for the identification of the membership functions, we can use methods in nonlinear programming to search for the optimal parameters since the performance index, i.e., the output error, is nonlinear with respect to the parameters of the membership functions. Here we have to iterate

(2) the identification process of the coefficients since those depend on (3) the membership functions in the premises. Also we have to iterate (2) and (3) for (1) each choice of a fuzzy partition of the input space. Fig. 1 shows the outline of the identification algorithm.

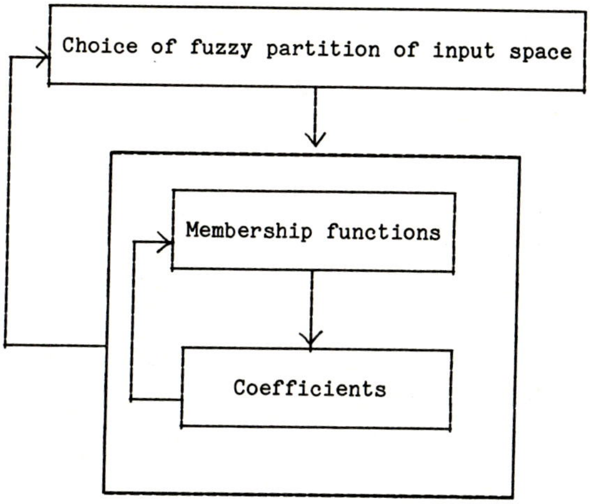

Fig. 1 Outline of identification algorithm

MODEL CAR

Control situation

The purpose of control in this study is to put a car into a garage as a man does. One of the authors has designed a fuzzy controller to navigate a car through a turn [11]. It has been shown that a car runs smoothly through a turn as if it were driven by an expert.

This study has been started as the next step following the former study.

SIDE WALL

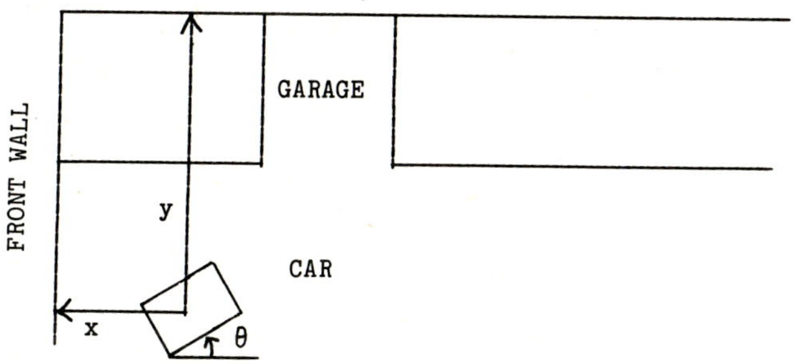

Fig. 2 Control situation

Fig. 2 shows a garage surrounded by two walls and a car. The space in front of the garage is too narrow to put a car directly into the garage. So one needs some techniques for parking the car. The walls are located so that a car measures its direction and position by the sensing device.

The following variables are chosen as input information to the fuzzy controller.

x : front wall distance
y : side wall distance
θ : heading angle

Configuration of the model car

Fig. 3 shows the appearance of the model car. It has the following configuration.

Body: length 40 cm, width 21 cm, height 30 cm, weight 3.5 kg
Drive force of rear wheels: DA motor
Steering control of front wheels: servo motor
Sensor: two supersonic sensors driven by a stepping motor for the measurements of distance and direction
Fuzzy controller: on board micro-processor, CPU Z80 with 4 KB ROM, 1 KB RAM and DA/AD converters
Batteries: 1.5 V x 8

Fig. 3 Appearance of the model car

We can see two supersonic sensors as well as a stepping motor on the top of the model car. As is seen at the configuration, the model car has the ability of self-running without any help from outside. It can take three discrete speeds: positive (forward), zero (stop) and negative (backward). These are constants: 4cm/sec. Fig. 4 shows the configuration of the hardware.

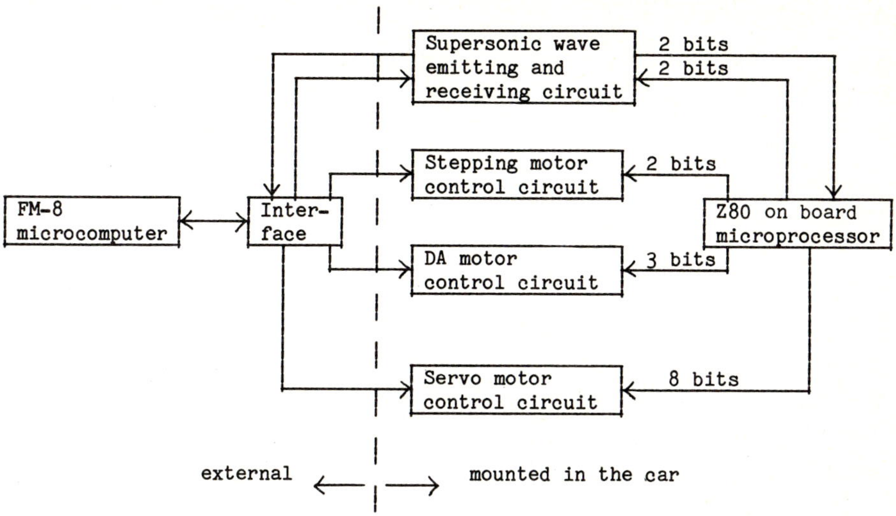

Fig. 4 Configuration of the hardware

Measurement of distance: x and y

A sensor is made by using 40 KHz supersonic transducer. It can measure distance by counting the time difference between the time of emitting a supersonic wave and that of receiving its echo reflected by the front or the side wall. The resolution of sensing is about 2 cm. For the sake of saving the sensing time, two sensors are put at right angles to each other; they can measure the both x and y at the same time.

Measurement of direction: θ

The supersonic sensors are rotated just like a radar by a stepping motor. When the sensing unit is parallel with the wall, it can recieve the supersonic echo. Since a supersonic wave has a good directivity, the direction of the car can be measured by counting the number of pulses put into a stepping motor while the sensing unit turns from the home position to the position of catching the echo. The stepping motor is rotated by 3.6° with 1 pulse. The resolution of the sensor is 7.2°.

Derivation of control rules

The model car can be driven by a man through a micro-computer, as is seen in Fig. 4, which is used to gather data. Fig. 5 shows the trajectories of the car when an trained person controls it.

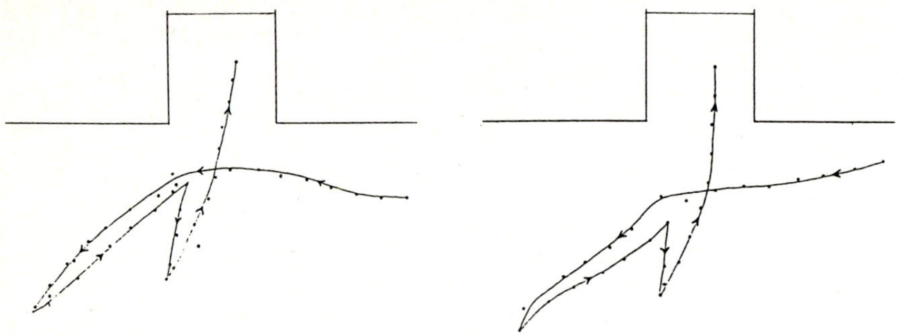

Fig. 5 Trajectories of the model car driven by a man

From six trajectories five variables are measured. Those are x, y, θ, and the following two:

f: angle of the front wheels in moving forward
b: angle of the front wheels in moving backward

These two variables are the outputs from the fuzzy controller. The number of data samples for identification is about 500. There is another output, the index s for speed control.

s: index to stop the car, $-10 \leqq s \leqq 10$

The control rules are classified into three groups: (1) the steering control in moving forward, (2) the steering control in moving backward and (3) the speed control. A control rule of the first group has the following structure.

If x is A, y is B, θ is C then $f = p_0 + p_1x + p_2y + p_3\theta$.

Fig. 6 shows some of the eighteen control rules for the steering control in moving forward.

Similarly the second group consists of the control rules for the steering control in moving forward where a fuzzy control rule is of the form:

If x is D, y is E, θ is F then $b = q_0 + q_1x + q_2y + q_3\theta$.

Fig. 7 shows some of the sixteen control rules.

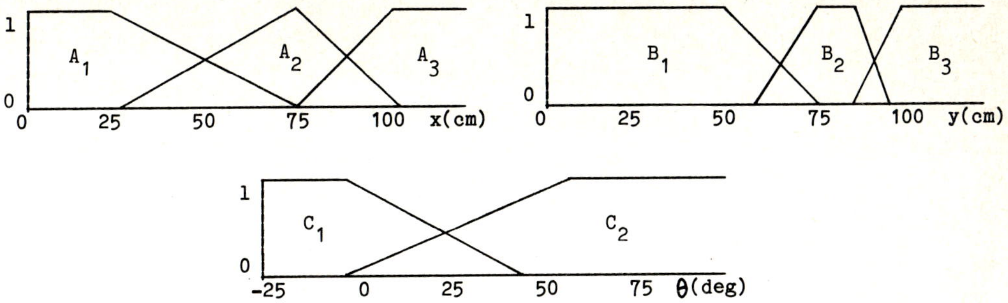

Fuzzy variables in the premises

rule number	x	y	θ	p_0	p_1	p_2	p_3
4	A1	B2	C2	997	5.23	-10.8	0.88
8	A2	B1	C2	-2310	13.2	29.9	-3.44
11	A2	B3	C1	720	-14.1	7.64	2.63
14	A3	B1	C2	996	11.3	-48.3	0.62
16	A3	B2	C2	2070	-3.45	-5.97	3.02

Coefficients in the consequents

Fig. 6 Control rules in moving forward

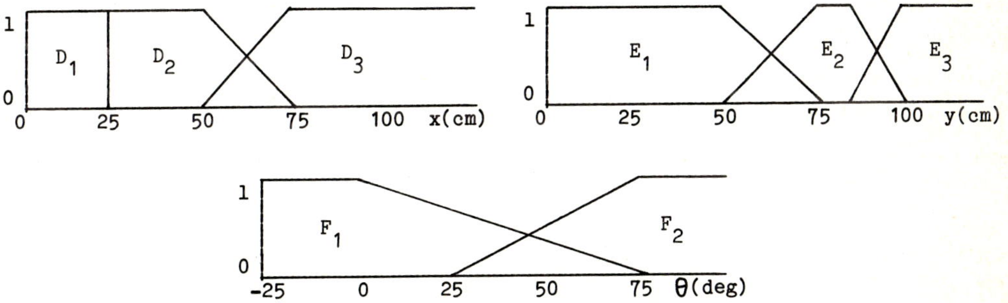

Fuzzy variables in the premises

rule number	x	y	θ	q_0	q_1	q_2	q_3
4	D1	E2	F2	2310	-10.9	-20.2	22.6
7	D2	E1	F1	3000	7.74	18.1	5.40
11	D2	E3	F1	-1090	2.50	5.14	-0.62
15	D3	E2	F1	-2150	1.76	-1.71	-2.72

Coefficients in the consequents

Fig. 7 Control rules in moving backward

Now the third group is for the speed control where the control rules output three values: -10, 0 and 10. In order to form fuzzy control rules giving these discrete values, we set up an artificial variable s(t), as a function of time. Fig. 8 shows the change of the variable according to the movement of the car.

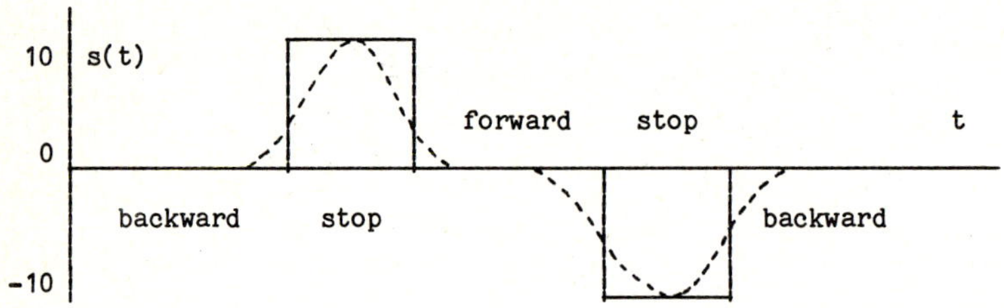

Fig. 8 Artificial variable for speed control

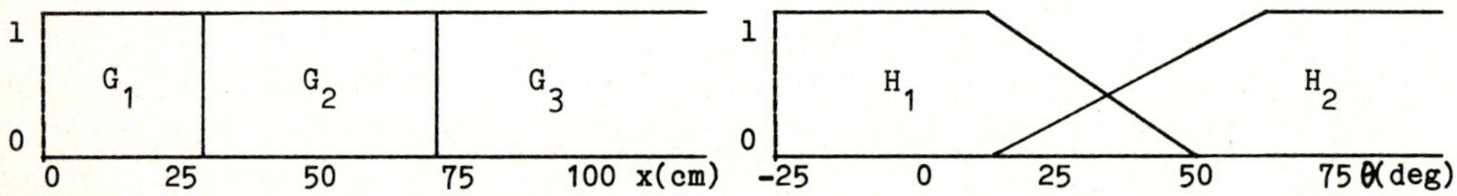

Fuzzy variables in the premises

rule number	x	θ	r_0	r_1	r_2	r_3	r_4
1	G1	H1	56.9	0.20	-0.21	0.05	-3.19
4	G2	H2	-84.1	63.2	0.18	0.03	-0.18
5	G3	H1	3.07	0.27	0.02	0.01	-12.4

Coefficients in the consequents

Fig. 9 Speed control rules

The structure of fuzzy control rules for speed control is chosen by observation of a man's way of driving a car as follows.

If x is G, θ is H then $s = r_0 + r_1 x + r_2 y + r_3 \theta + r_4 (f \text{ or } b)$

where f and b in the consequent are the outputs from the control rules of the first group and those of the second group, respectively. The variable f is used in the consequent when the car is moving forward and b is used when it is moving backward. As is seen from the format of a control rule, the variable y, i.e., the side wall distance does not appear in the premises.

The parameters of the rules are identified so that the output of the rules coincides with the stepwise function s(t). Fig. 9 shows the rules for speed control.

The actual output of the control rules is shown by a dotted line in Fig. 8. So the change of the speed is performed by setting a threshold.

EXPERIMENTS

We use an on board micro-processor with ROM and RAM as a fuzzy controller which is mounted on the model car. The fuzzy control program with 40 control rules including subprograms for measurements and I/O interfaces is printed in ROM. The size of the program is 4KB in machine language. Fig. 10 shows the flowchart of the control program.

Some results of fuzzy control are shown at Fig. 11 (a) - (d). Among those (c) and (d) show failure to park the car. The failure is caused by sensing error and the defects of speed control. As is discussed previously, the speed is also fuzzily controlled by 6 control rules. However it seems better to change the speed under crisp conditions concerning the position of the car. As a whole, it has been found that the fuzzy controller has a good performance.

CONCLUSIONS

This paper has discussed a method for deriving fuzzy control rules from operator's control actions. The method has been applied for the design of a fuzzy controller for parking a model car where a fuzzy controller is realized in on board microprocessor. It has been shown that the presented method works very well.

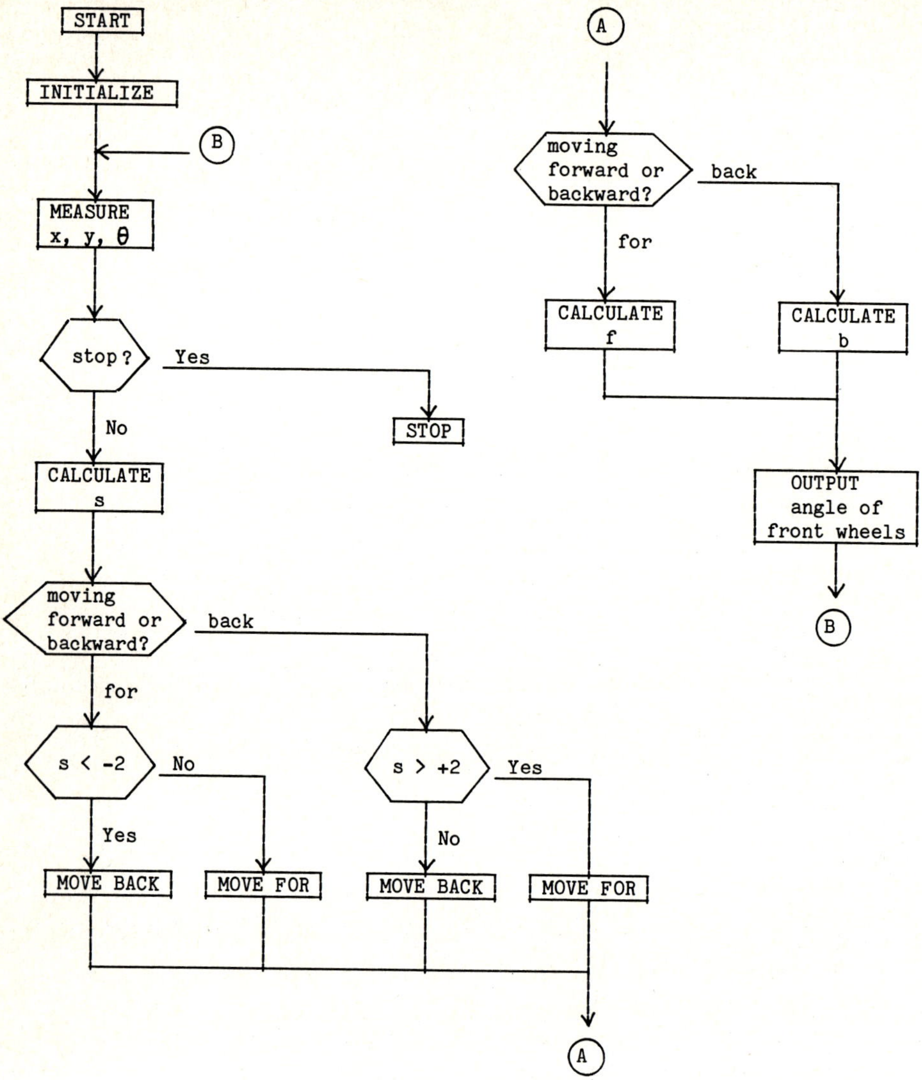

Fig. 10 Flowchart of control program

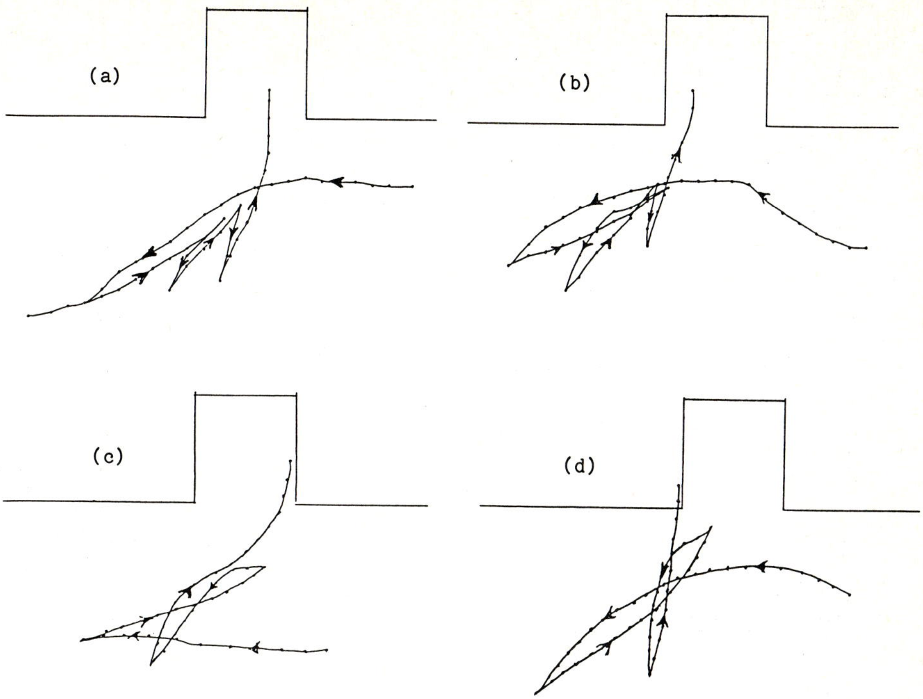

Fig. 11 Results of fuzzy control

REFERENCES

[1] E. H. Mamdami: Applications of Fuzzy Algorithms for Control of Simple Dynamic Plant, Proc. IEE, 121, 12, 1585-1588, 1974

[2] D. Willaeys, N. Malvache, P. Hammad: Utilization of Fuzzy Sets for System Modelling and Control, Proc. 16th IEEE Conf. Decision and Control, New Orleans, December 1977

[3] R. M. Tong: Analysis and Control of Fuzzy Systems Using Finite Discrete Relations, Int. J. General Systems, 4, 143-162, 1978

[4] M. Braae, D. A. Rutherford: Theoretical and Linguistic Aspects of the Fuzzy Logic Controller, Automatica, 15, 5, 553-577, 1979

[5] L. P. Holmblad, J. J. Ostergaard: Control of a Cement Kiln by Fuzzy Logic, in: Fuzzy Information and Decision Processes, M. M. Gupta, E. Sanchez (eds.), North-Holland, Amsterdam, 1982

[6] E. Czogala, W. Pedrycz: Fuzzy Rule Generation for Fuzzy Control, Cybernetics and Systems, 13, 3, 275-294, 1982

[7] T. Takagi, M. Sugeno: Derivation of Fuzzy Control Rules from Human Operator's Control Actions, Proc. IFAC Symp. on Fuzzy Information, Knowledge Representation and Decision Analysis, Marseille, July 1983

[8] M. Sugeno, T. Takagi: A New Approach to Design of Fuzzy Controller, in: Advances in Fuzzy Sets, Possibility Theory, and Applications, P. P. Wang, S. K. Cheng (eds.), Plenum Press, New York, 1983

[9] M. Miyamoto, S. Yasunobu: Predictive Fuzzy Control and its Application to Automatic Train Operation Systems, First International Conference on Fuzzy Information Processing, Hawaii, July 1984, also in this volume.

[10] O. Yagishita, O. Itoh, M. Sugeno: Application of Fuzzy Reasoning to Water Purification Process, in this volume.

[11] M. Sugeno, M. Nishida: Fuzzy Control of Model Car, Fuzzy Sets and Systems, to appear.

Industrial Applications of Fuzzy Control
M. Sugeno (ed.)

A FUZZY CONTROLLER IN TURNING PROCESS AUTOMATION

Y. Sakai and K. Ohkusa

Department of Mechanical Engineering
Yamaguchi University
Ube, Yamaguchi
Japan

Basic ideas are described, which are necessary in dealing with automatic regulation of the cutting status. A human operator's way and its characteristics in manual handling are analysed, and their fuzzy theoretic translation is tried showing some concrete examples.

INTRODUCTION

The recent remarkable development in electronics also brings many very useful tools to the field of machining, and therefore new ideas are being brought about in this field. But as is often the case, there are still many procedures in machining which are rather hard to replace by electromechanical devices. As an example, the evaluation of the quality of a product has always been performed by human inspection. Lately, however there have been attempts to automate this human labour. The difficulty in so doing resides in the fact that the measurement and scaling should be multi-dimensional and are subjective, sometimes including psychological factors. So we do not know what kinds, nor how many, physical quantities should be chosen for the substitution.

To adjust the cutting condition, it is necessary to know the situation as precisely as possible. But at the same time it is difficult to do so as mentioned above. Incorporation of human factors will provide a successful solution to this problem.

TASK OF THE PRESENT CONTROLLER

The so called FMS is now one of the major subjects of discussion and has already been installed in many manufacturing industries.[1] There are still various problems unsolved. As mentioned above, the bottleneck is human factors.

The purpose of the present controller is the recovery of the cutting status in turning operation when failure occurs. What is meant by the word failure are a bad condition in chip generation (undesirable chip flow), chatter, and tool fracture. This methodology can be easily extended to any stock removal machining processes. The automatic and adaptive regulation of the cutting status is one of the desired tasks surrounding machining.

Machining itself has been computerized and this has brought benefits to practical manufacturing works. But such a system performs just a pre-programmed task irrespective of the present cutting situation. To prevent troubles, and at their occurrence, human intervention is necessary. This limits the progress of automatization; details can be found in reference [2]. But this is reasonable in one aspect, because the mechanism of machining is not clarified yet. As an example, the chemical compositions of the work and tool materials employed greatly affect machinability. But a unified explanation of this mechanism is hard to make. Experimental work for the specific case is the only available way for finding the optimal operating condition. The idea of a database for machining operations exists and is applied in practical situations for the initialization of operating conditions. But there is difficulty in the application of this idea.

For instance, different machine tools or their different historical status require different condition settings. Only the human operators, especially skilled operators can overcome this difficulty. They can deal with the situations adaptively. From this view point, the authors are attempting to introduce such ability of a human being in establishing the function of automated supervision. The most feasible clue to this question is the idea of fuzzy control. [3, 4, 5] The following example is a direct application of the methodology.

Table 1 shows the control strategies in a linguistic form. Each of them is an implication to recover the cutting status in an undesirable situation in chip generation. Each takes the following form:

If P (the status in chip generation),
then Q (increment of feed rate to be given for adjustment). ... (1)

where P and Q are statements. Each of them is provided verbally, so both the antecedent and consequent are fuzzy in the sense of fuzzy set theory. In other words, the predicates of the statements do not indicate a definite value but instead they include more or less vagueness. Besides the terms which appear in the implication are subjective, and hence the interpretation should be given. They are shown in Tables 2-a and 2-b.

An auxiliary control action can be added to the primary action cited, if necessary.

In this case, the implications can be written as:

If P then Q and R.

where R indicates the additional action to be taken. R's as change in cutting speed are also shown in the lower part of Table 1, and the corresponding meaning of the fuzzy subsets used are shown in Table 2-c. The statements P and Q (or P and R) form a fuzzy binary relation. (The adjective "binary" will be omitted in the sequel.) A fuzzy relation R between fuzzy subsets $A \subset C$ and $B \subset X$ is a subset of the Cartesian plane C x X, and is characterized by the following membership function:

$$\mu_R(c, x) = \mu_A(c) \wedge \mu_B(x), \qquad \ldots (3)$$

where $c \in C$ and $x \in X$. In the case of discretized coordinates into finite sections as in the present case cited in the tables, the membership function of a relation can be written in a matrix form with the (i, j) elements:

$$\mu_{ij} = \mu_R(c_i, x_j) = \mu_A(c_i) \wedge \mu_B(x_j), \qquad \ldots (4)$$

where c_i and x_j are the i-th smallest element of the support of A and the j-th smallest element of the support of B, respectively; the support of a fuzzy set is the nonfuzzy set where the corresponding fuzzy set has nonzero grade of membership. In the above definition of a fuzzy relation, the extent of the relationship is determined by the lowest grade of membership of those of the corresponding elements c and x which in turn is the highest degree satisfied by both c and x.

The instructions are linguistic and ambiguous. But after all a definite value of the increment to be chosen is given through the calculational procedure. So finally the procedure becomes deterministic. This is similar to a human operator's procedure for his decision-making and action in accordance with his decision. As a familiar example, consider a case of driving a car on a curve in a road (as a part of circle, for simplicity). He knows that he must turn his steering wheel "this much," but he can not tell a definite value of the angle to be steered to go through the corner. Generally the width of the road is sufficiently wide to pass through by an automobile, and any value is applicable if it clears the constraint given by the road. He knows the critical angles "fuzzily." But actually he turns his steering wheel by a single definite value almost unconsciously. Every time he passes through the same curve driving the same car, the magnitude of the angle is possibly different, almost the same though. (This is a statistical or probabilistic feature of such actions.) So the present procedure can be considered as just a duplication of a human operator's way of control, and automating this procedure can only perform the "pre-programmed" task.

The application of the instructions provided gives results as shown in Table 4. Table 3 is convenient for finding the action required. The original status of chip generation was "snarled." The first regulation was made in accordance with the instructions as given in Table 3. The result obtained was slightly insufficient. Hence again adjustment was tried successfully. Originally each implication is so selected that it is sufficient to recover the chip status. This discrepancy lies in:

1) The instructions are given based on the empirically acquired knowledge.
2) The relationship between very different situations are difficult to comprehend in terms of operating conditions.

The latter is especially worth noting. In this respect an idea will be discussed later.

Table 1 *Control Rules for Chip Control*

	Antecedent (Chip Length)	Consequent (Required Change in Feed Rate)
1	Snarled	Positive Very Big
2	Excessively Long	Positive Big
3	Very Long	Positive Medium
4	Long	Positive Small
5	(All Right)	(Zero)
6	Short	Negative Small
7	Excessively Short	Negative Medium

[Auxiliary Adjustment by Changing Cutting Speed]

	Change in Cutting Speed
1	Negative Big
2	Negative Medium
3	Negative Small
4	Negative Small
6	Positive Small
7	Positive Medium

Table 2 *Translation of Linguistic Terms into Fuzzy Subsets*

Table 2-a

	Classification of Chip Length						x 1 mm
	∿15	15∿25	25∿35	35∿45	45∿55	55∿	Continuous
Snarled	0	0	0	0	0	0.3	1.0
Excessively Long	0	0	0	0	0.5	1.0	0.3
Very Long	0	0	0	0.5	1.0	0.5	0
Long	0	0	0.3	1.0	0.5	0	0
All Right	0	0.3	1.0	0.3	0	0	0
Short	0.3	1.0	0.3	0	0	0	0
Excessively Short	1.0	0.3	0	0	0	0	0

Table 2-b

	Change in Feed Rate						x 0.01 mm/rev
	-2	-1	0	1	2	3	4
Positive Very Big	0	0	0	0	0	0.3	1.0
Positive Big	0	0	0	0	0.5	1.0	0.3
Positive Medium	0	0	0	0.5	1.0	0.5	0
Positive Small	0	0	0.3	1.0	0.3	0	0
Zero	0	0	1.0	0	0	0	0
Negative Small	0.3	1.0	0.3	0	0	0	0
Negative Medium	1.0	0.3	0	0	0	0	0

Table 2-c

	Change in Cutting Speed					x 10 m/min
	-3	-2	-1	0	1	2
Positive Medium	0	0	0	0	0.3	1.0
Positive Small	0	0	0	0.3	1.0	0.3
Zero	0	0	0	1.0	0	0
Negative Small	0	0.5	1.0	0.3	0	0
Negative Medium	0.3	1.0	0.5	0	0	0
Negative Big	1.0	0.3	0	0	0	0

Table 3 *Control Actions Obtained*

	Classification of Chip Length						
	∿15	15∿25	25∿35	35∿45	45∿55	55∿	Continuous
Change in Feed Rate	-2	-1	0	1	2	3	4 x 0.01 mm/rev
Change in Cutting Speed	2	1	0	-1	-1	-2	-3 x 10 m/min

Table 4 *An Example of Practiced Results*

Number of Iteration	Chip Length	Change in Feed Rate	Change in Cutting Speed	Cutting Condition
0	Continuous	4×10^{-2} mm/rev	Not Applied	Speed = 140 m/min Feed = 0.15 mm/rev Cut Depth = 2 mm
1	35∿45	1×10^{-2} mm/rev	Not Applied	140 0.19 2
2	25∿35	Finished		140 0.20 2

If the value of change recommended by a control rule is too big to compensate for the deviation in cutting status, then the regulation action taken may result in a worsened situation. So the rules should be determined carefully. If, on the contrary, the indicated value by the rule is insufficient for the adjustment, the matter may be better, since by that action the situation is more or less improved, and iteration of control action may be more successful than in the former case. Anyway the question of how to obtain good performance without instability is a subject to be solved, but as implied in the above it is a heuristic way to provide control rules so that the control actions are moderate.

The interpretation of verbal instructions, say, finding (the fuzzy set of) real number(s) which corresponds to "positive big," can be done in two ways: one is the direct translation by the human operator who gives the rules, and the other is the one determined subjectively to the software who receives the instructions. The latter is natural and common practice in a discourse. But as the primary step in the development of the supervisory system, the former is the easiest method introduced.

REVISED METHOD FOR PROVIDING CONTROL STRATEGIES

If the control strategies are selected more carefully, there may not be such an improper situation as that in the previous section. But it is natural in one aspect, because we human beings are accustomed to thinking of things in a linear manner, and nonlinearity is a cause of confusion. Looking at the situation from this point of view, a revised way for providing strategies can be found. It may be impossible to understand globally the relationship between the operating condition and the realized corresponding cutting status. But it is possible in any case to grasp it locally as a linear relation. Taking the union of all such local grasps in a manner described in what follows, we can guess the global relationship.

Besides the basic instructions provided this way by a human operator is not absolutely true. So the software should have his own intelligence for better control action.

To begin with, we must define the cutting condition space and the cutting state space. The cutting condition is a set of operating variables which includes feed rate, cutting speed, depth of cut and so forth. This ordered tuple is a point in the n-dimensional Euclidean space, the n-fold product of the set of all the real numbers. The factors other than the variables listed above may be discrete parameters in denoting the cutting condition space. So basically,

$$X_c = \{x: x = (x_1, \ldots, x_n)^T, x \in R^n\}, \qquad \ldots (5)$$

and if it is necessary to indicate discrete parameters explicitly,

$$X_c = X_c(p_1, \ldots, p_r), \qquad \ldots (6)$$

where $p_1, \ldots, p_r$ are discrete parameters such as tool material.

A cutting state variable s_i is the representation of a feature of the cutting situation in operation through its evaluation as a point in the real axis:

$$s_i = f_i(x_1, \ldots, x_n; p_1, \ldots, p_r), \qquad f_i: X_c \rightarrow R. \qquad \ldots (7)$$

A cutting state s is a vector or say, m-tuple:

$$s = (s_1, \ldots, s_m)^T \in R^m, \qquad \ldots (8)$$

or

$$s = f(x) = (f_1(x), \ldots, f_m(x))^T. \qquad \ldots (9)$$

The entire set of all such s's will be called the cutting state space S:

$$S = \{s: s = (s_1, \ldots, s_m)^T\}.$$

As the above notation shows, s is uniquely determined by choosing a cutting condition x (unique at least locally except at finite singular points). What are the cutting state variables depends on what features of cutting status one pays attention to, and on how precisely he wants to represent or evaluate the cutting status.

The cutting status is essentially an abstract entity. The cutting state is its representation in the m-fold product of R, and X_c plays the coordinate map as in Figure 1. If the mapping

$$f: X_c \rightarrow S, \qquad f = (f_1, \ldots, f_m)^T, \qquad \ldots (10)$$

is analytic, then for any $x_0 \in X_c$, in some neighborhood $N(x_0)$ of x_0 the following equation holds as a linear approximation:

$$ds = F(x_0)dx, \qquad \ldots (11)$$

where F(x) is the Jacobian matrix at x, and has the form of

$$F(x) = [\frac{\partial f_i(x)}{\partial x_j}],$$

and

$$s_0 = f(x_0), \qquad s \in N_s(s_0) - \{s: s = f(x), x \in N(x_0)\}.$$

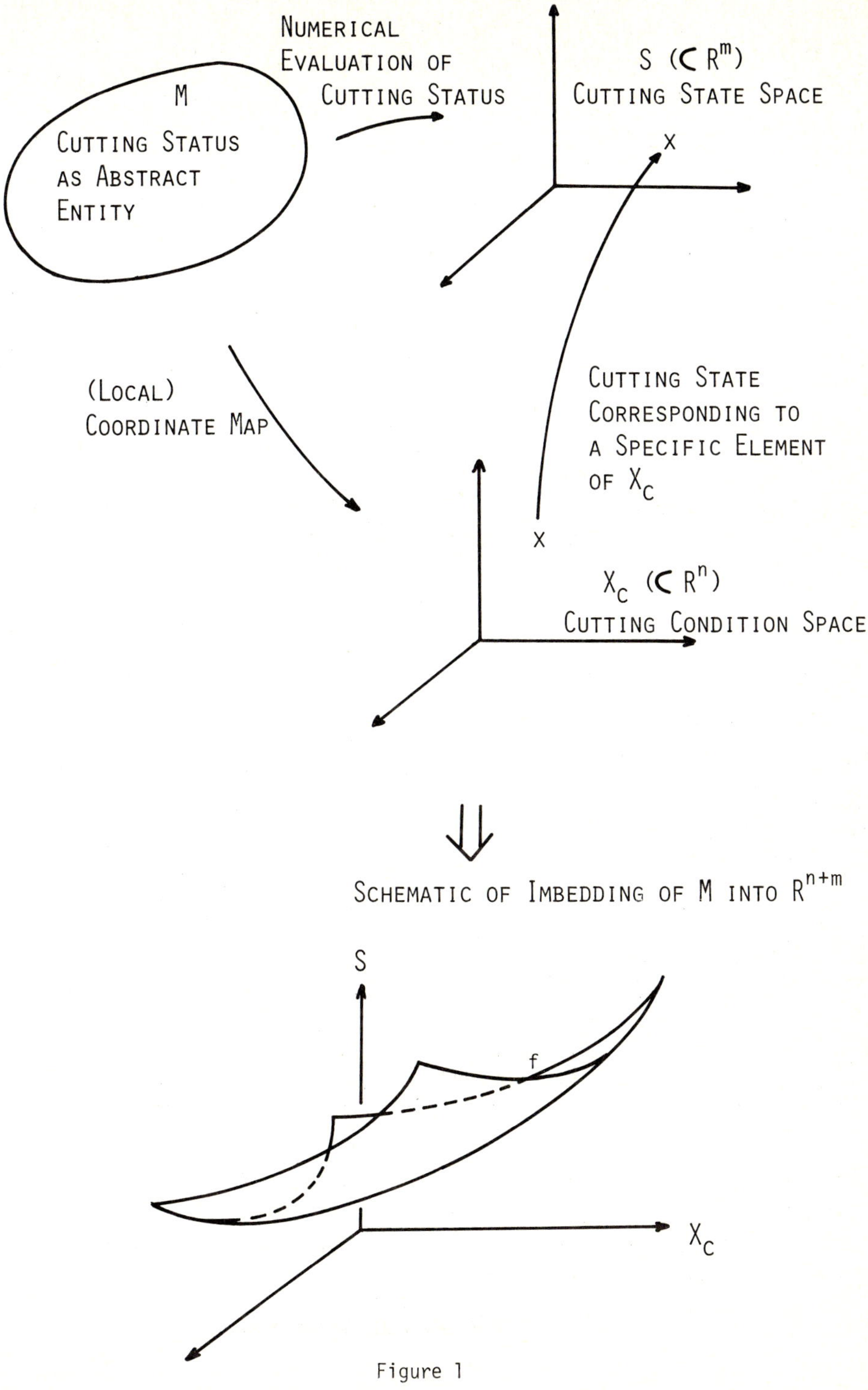
NUMERICAL
EVALUATION OF
CUTTING STATUS
M
CUTTING STATUS
AS ABSTRACT
ENTITY
S (⊂ R^m)
CUTTING STATE SPACE
X
(LOCAL)
COORDINATE MAP
CUTTING STATE
CORRESPONDING TO
A SPECIFIC ELEMENT
OF X_C
X
X_C (⊂ R^n)
CUTTING CONDITION SPACE
SCHEMATIC OF IMBEDDING OF M INTO R^n+m
S
f
X_C

Figure 1

The above deterministic description for $X_c \times S$ is the origin of the present idea for the refinement of providing control strategies and it can be readily translated into the fuzzy theoretic formulation. A neighborhood and its image in S can be fuzzified together with the grade of membership at every point in them. For the fuzzy case, instead of equation (11) we obtain, formally: (Fuzzy variables (subsets) will not be distinguished in symbol from the ordinary ones.)

$$ds = dx \circ R_{N(x_0)},$$

or ... (12)

$$(ds_1,\ldots, ds_m) = (dx_1,\ldots, dx_n) \circ \begin{pmatrix} R_{11} & \cdots & R_{1m} \\ \vdots & & \\ R_{n1} & \cdots & R_{nm} \end{pmatrix}.$$

More specifically,

$$ds_i = \sum_j dx_j \circ R_{ji}. \qquad \ldots (13)$$

The summation sign $\sum$ indicates "fuzzy addition" and needs a definite definition, but practically it can be understood as follows. Letting

$$ds_{ij} \triangleq dx_j \circ R_{ji}, \qquad \ldots (14)$$

ds_i can be represented or implied by

$$ds_i^0 = \Sigma_j ds_{ij}^0, \qquad \ldots (15)$$

where ds_{ij}^0 is an element of ds_{ij} with the maximal grade of membership corresponding to dx_j. Equation (15) is an analogy of an ordinary linear equation

$$ds_{ij} = a_{ij} dx_j, \qquad \ldots (16)$$

where a_{ij} is a constant real number. In this sense, R_{ji} can be regarded as the union of relations

$$R_{ji} = \bigcup_k R_{ji}^k, \qquad \ldots (17)$$

R_{ij}^k's being arranged in a "linear" manner in the Cartesian plane, $X_j \times S_i$.

In practical situations, only dominant R_{ji}'s are employed in describing $R_{N(x)}$, and the remaining R_{ji}'s which are rather less effective in adjusting cutting state are omitted and set zero. Through this procedure $R_{N(x)}$ becomes less accurate as an approximation of F(x). Further one usually does not understand equation (11) or (12) in vector form as given. Rather, one can understand it componentwise or as a scalar correspondence. Hence, although it is just formal, the following inverse form of equation (12) is applicable:

$$dx_j = ds_i \circ R_{N(x)}^{-1}. \qquad \ldots (18)$$

MISCELLANEOUS PROCEDURES IN CONTROL

For adaptive machining control, the controller's own knowledge is necessary. It can be classified into:

1) Common sense – includes general education, intuition, and heuristics.
2) Special knowledge of machining – engineering knowledge.
3) Instructions and information provided by a human expert, the boss – includes the capacity of the machine tool employed.

Common sense which constitutes the elementary knowledge of control is useful, for example, in finding out careless mistakes in the expert's initialization of the operating conditions, and is convenient for quick information retrieval, serving as something like indices. A cutting speed about 80 m/min is a high speed for a HSS tool, but is rather low for a cemented carbide tool. As this example shows, "high cutting speed" has a variety of meanings. In other words, its meaning depends on the knowledge level. This fact also implies the necessity of knowledge ranking. Knowledge can be implicative statements with respect to the relationship between the cutting condition and the corresponding cutting state, or can be simply a fuzzy set in the real axis. The last one of level 3 in the above listing should be distinguished from the rest in that the strategies are defined in $\{(ds, dx)\}$, i.e., in terms of increments ds and dx. Note that there are three levels of undesirable situations:

a) Not O.K. for machining, but the machining is continuable as it is.
b) Not O.K. at all, but recovery can be expected by some regulation.
c) No use expecting recovery, or not having any effective regulation recipe.

Hence the above principle in providing instructions is applicable only to strategies which fall under the level b).

As an example of making use of knowledge 1), suppose that the following cutting condition is inputted by the boss in turning steel with a cemented carbide tool:

Feed rate $f = 0.4$ mm/rev,
Cutting speed $v = 200$ m/min,
Depth of cut $d = 3$ mm.

Referring to the controller's intuitive knowledge in Table 5, he finds that the pair (200 m/min, 0.4 mm/rev) has the grade of membership of 0.3. It seems slightly dangerous to him. Projection of the membership function into each set of cutting speed and feed rate (The projection operator will be described later.) gives,

$$\text{dangerous speed} = 0.3/180 + 0.5/190 + 0.8/200 + 1/210,$$
$$\text{dangerous feed} = 0.3/0.4 + 0.6/0.5 + 0.7/0.6 + 1/0.7,$$

and he finds that the predominant factor is the speed. Then he looks up the generally recommended range of cutting speed for turning steel with a cemented carbide tool around the feed rate of 0.4 mm/rev in his file of knowledge of level

Table 5 *Too High Cutting Speed and Too Big Feed Rate to be Applied (Knowledge of Level 1))*

(Feed Rate)	Grade of Membership			
0.7 mm/rev	0.3	0.5	0.8	1
0.6	0.3	0.5	0.7	0.7
0.5	0.3	0.5	0.6	0.6
0.4	0.3	0.3	0.3	0.3
	180	190	200	210 (Speed) m/min

2). Let the recommended range be 100 ∿ 160 m/min (This range is indicated fuzzily, but the question is put aside here.) for the feed of 0.4 mm for turning steel, then the speed of 200 m/min is found to be too high. (Evaluation of this fact can be made by taking the complement of the fuzzified recommended range of speed.) This result is worth giving a warning to his boss. But there may be a certain special reason in selecting this condition. In that case, it is necessary to estimate the required power for the operation at this condition in order to check the applicability of this condition to the lathe employed. (The data is a part of information of level 3).)

Fuzzy knowledge can also be made use of for many other purposes together with its membership function for a criterion function and with the aid of fuzzy operators. Various operators can be defined. A few of them will be described drawing our attention to the fact:

$$1 \wedge \mu = \mu,$$

where $\mu \in [0, 1]$ is a value of any membership function. For a relation,

$$R = \begin{pmatrix} 0 & 0 & 0 & 0 & 0 \\ 0 & 0.3 & 0.3 & 0.3 & 0 \\ 0 & 0.5 & 1 & 0.4 & 0 \\ 0 & 0.2 & 0.2 & 0.2 & 0 \\ 0 & 0 & 0 & 0 & 0 \end{pmatrix},$$

which is the relation of (0 0.3 1 0.2 0) to (0 0.5 1 0.4 0), we have the following concrete examples of operations:

$$\begin{pmatrix} 0 & 1 & 0 & 0 & 0 \\ 0 & 0 & 1 & 0 & 0 \\ 0 & 0 & 0 & 1 & 0 \\ 0 & 0 & 0 & 0 & 1 \\ 0 & 0 & 0 & 0 & 0 \end{pmatrix} \circ R = \begin{pmatrix} 0 & 0.3 & 0.3 & 0.3 & 0 \\ 0 & 0.5 & 1 & 0.4 & 0 \\ 0 & 0.2 & 0.2 & 0.2 & 0 \\ 0 & 0 & 0 & 0 & 0 \\ 0 & 0 & 0 & 0 & 0 \end{pmatrix}, \qquad \dots (19)$$

and

$$(1 \quad 1 \quad 1 \quad 1 \quad 1) \circ R = (0 \quad 0.5 \quad 1 \quad 0.4 \quad 0). \qquad \dots (20)$$

The former operation can be called shift operation and the latter, Projection. A discrete relation is decomposable into the original subsets which form the particular relation, if it is a normal relation (with the maximal grade of membership of unity). (Convexity of each subset and uniqueness of the maximal grade are also assumed.) Refer to Wenstøp [6] for the original idea of the former operator and to Kaufmann [7] for the definition of projection. The definition and application of fuzzy relations can be found in references [7, 8]. Problems which surround the automation of the supervision of cutting operations are described in references [9, 10].

REFERENCES

[1] Hartley, J., FMS at Work, co-published by IFS and North-Holland, 1984.

[2] Novak, A., "The Concept of Artificial Intelligence in Unmanned Production — Application of Adaptive Control," Proceedings of the 2nd International Conference on Flexible Manufacturing Systems, 26-28 Oct. 1983, London, U.K., pp.669-680.

[3] Mamdani, E.H., "Application of Fuzzy Logic to Approximate Reasoning Using Linguistic Synthesis," IEEE Trans. on Computers, Vol.C-26, No.12, Dec. 1977, pp. 1182-1191.

[4] Kickert, W.J.M., and Mamdani, E.H., "Analysis of a Fuzzy Logic Controller," Fuzzy Sets and Systems, 1, 1978, pp.29-44.

[5] Mamdani, E.H., and Gaines, B.R., Eds., Fuzzy Reasoning and its Applications, Academic Press, 1981.

[6] Wenstøp, F., "Deductive Verbal Models of Organizations," ibid., pp.149-167.

[7] Kaufmann, A., Introduction to the Theory of Fuzzy Subsets, I, Academic, 1975.

[8] Zadeh, L.A., "Outline of a New Approach to the Analysis of Complex Systems and Decision Processes," IEEE Trans. on Systems, Man, and Cybernetics, Vol.SMC-3, No.1, Jan. 1973, pp.28-44.

[9] Sakai, Y., Nakato, T., and Ohkusa, K., "Identification and Failure - Detection System for Cutting Process," Proc. IMEKO TC-14 International Symposium on Metrology for Quality Control in Production, 10-13 July, 1984, Tokyo, Japan, pp.62-67.

[10] Sakai, Y., and Ohkusa, K., "On a Control System for Cutting Process," Proc. 16th CIRP International Seminar on Manufacturing Systems, 13-14 July, 1984, Tokyo, Japan, pp.188-195.

Industrial Applications of Fuzzy Control
M. Sugeno (ed.)

DESIGN OF FUZZY CONTROL ALGORITHMS WITH THE AID OF FUZZY MODELS

Witold Pedrycz

Department of Automatic Control and Computer Science
Silesian Technical University
44-100 Gliwice, Pstrowskiego 16
Poland

Fuzzy control algorithms have been studied as a formal approach to process control while we have to cope with fuzziness present in the system studied. Usually a fuzzy controller is viewed as the control algorithm based on a set of control rules creating a linguistic protocol of actions performed by a human being. A significant number of different ways of inference of fuzzy control for the linguistic protocol and available state of the system have been put into extensive studies. In this paper we deal with designing fuzzy control algorithms on the basis of a fuzzy model of the system. The notion of the fuzzy model used here plays with linguistic labels and linguistic relationships between them. A method of construction of the relation of the controller is described in detail. Moreover a controllability index is also introduced and discussed. Identification and control tasks relevant to fuzzy systems are solved in a framework of fuzzy relation equations. The proposed design procedures of fuzzy control are illustrated by means of real industrial data of gas furnace.

Key words : fuzzy control, fuzzy relation equation, fuzzy controller, controllability

INTRODUCTION

Fuzzy control [4][7][21] started in the 70's has been studied extensively for practical control of systems in which we have to cope with fuzziness. In process control one may clearly realize some evident sources of fuzziness. Let us recall two of them,

a. fuzziness concerned with the system description. It is mainly due to lack of knowledge of the system, e.g. one cannot specify accurate dependencies between input (control) and state variables (e.g. in form of any function obtained by regression analysis), sources and character of distortions associated with the system.

b. fuzziness concerned with goals and constraints in the system controlled. Both of them may be specified in a linguistic manner, e.g. high incomes, low fuel consumption etc. There is no reason to specify exactly one numerical value as the appropriate representation of the goal or constraint. Both the goals and constraints are combined together conjunctively.

The above mentioned sources of fuzziness, on the one hand, and available control protocols on the other, cf. [15] form direct reasons for the development of fuzzy controllers. Similarly as stated in [17] three modes of derivation of fuzzy control rules are distinguished,

-based on operator's experience and/or control engineer's knowledge,

-based on the fuzzy model of the process controlled,

-based on operator's control actions.

The first method is commonly used, cf. [4][7]. It consists of two main steps: (i) knowledge acquisition collecting control rules, (ii) combining control rules and inference mechanism. In (i) we collect knowledge on control of the system, usually the rules take the form of "if-then" statements, where antecedents and consequent of every implication contain fuzzy data. The second step is of more delicate nature and works by combining all the statements together in the form of a fuzzy relation and the infering control for any condition provided.

Usually one applies the compositional rule of inference [22] or some implementations of the modus ponens in the field of fuzzy reasoning and fuzzy logic, cf. [1][7][8][19]. For another point of view on reasoning with fuzzy data see also [14]. While following this way we have to answer a few questions that are of interest in direct applications of fuzzy controllers,

-how we can learn the set of control rules is sufficient, noncontradictory, and noninteractive,

-which mechanism of inference should be used in a concrete system,

-how to measure quality of the fuzzy controller designed.

All of them are not exhaustively discussed until now. In case studies the fuzzy controllers work well and possess nice dynamical properties caused by the nonlinearity of the controller, cf. [4]. They contain a control policy that is insensitive with respect to random disturbances. Nevertheless we have to remember that the achievement of such nice properties has been attained by trial-and-error optimization. Only few self-learning procedures have been studied [9].

The second way is based on calculation of control actions while the fuzzy model of the system is provided. The idea behind this way is not a new one and it is well known in control engineering. The main point lies in the fact that we have to construct the fuzzy model of the system. A main difference between the models that are studied in control engineering and fuzzy models lies in the character of the quantities that are considered. In fuzzy models we deal with linguistic labels and relationships between them, while in the models of control theory we handle numerical values. The above mentioned two ways are also distinct in the sense of handling the performance index under optimization,

- while collecting control rules we do not know the goal of control in an explicit manner. It is "coded" in the control rules themselves,
- in the second approach we have to specify the goal and constraints, maybe fuzzy, and derive fuzzy control.

In the paper we will follow this second way presenting firstly a construction of fuzzy models described with the aid of fuzzy relation equations, and secondly discuss the solution of control problems, that in turn leads to the construction of the fuzzy controller. We will introduce a notion of controllability of the fuzzy system, as well. Numerical considerations are devoted to the presentation of fuzzy control algorithms for gas furnace data.

FUZZY MODEL DESCRIPTION AND FUZZY SYSTEM IDENTIFICATION

Here we focus our attention on fuzzy models of nondeterministic systems that are described by means of fuzzy relation equation,

$$X_{k+1} = U_k \square X_k \square R, \tag{1}$$

that models the system in which fuzziness is joint with its structure a fuzzy relation R and state, and control, X_k, U_k fuzzy sets. More precisely, X_k, X_{k+1}, U_k are fuzzy sets defined in the space of state and control, respectively,

$$X_k,\ X_{k+1} : \mathbf{X} \rightarrow [0,1],\ U_k : \mathbf{U} \rightarrow [0,1].$$

R denotes the fuzzy relation given in cartesian product $\mathbf{U} \times \mathbf{X} \times \mathbf{X}$, viz. $R : \mathbf{U} \times \mathbf{X} \times \mathbf{X} \rightarrow [0, 1]$. Further on we assume $\mathbf{U}$ and $\mathbf{X}$ are finite, that completely sufficies for applications presented in Section 4, or in general is appropriate for any computer implementation of the control algorithms studied. "$\square$" is sup-t composition [9] where "t" denotes triangular norm [16], viz. associative, monotonic, commutative function of two variables, $t : [0, 1] \times [0, 1] \rightarrow [0, 1]$ fulfilling boundary conditions, $a t 0 = 0$, $a t 1 = a$, $a \in [0, 1]$. The membership function of X_{k+1} standing in (1) is read accordingly,

$$X_{k+1}(y) = \max_{u \in \mathbf{U}, x \in \mathbf{X}} [U_k(u)\ t\ X_k(x)\ t\ R(u, x, y)], \quad y \in \mathbf{X}. \qquad (2)$$

Analogously, as in dynamic systems described by difference equations, we may modelize fuzzy systems of higher order, for instance the p-th order. Now the state in the (k + p)-th discrete time moment depends on past history states of the system at a certain horizon being equal to "p" and the fuzzy control,

$$X_{k+p} = U_k \square X_k \square X_{k-1} \square \ldots \square X_{k-p} \square R \qquad (3)$$

Now we prove the following,

Proposition 1.

The fuzzy relation model (3) of the p-th order can be reduced to the fuzzy relation model of the first order, such as,

$$\tilde{X}_{k+1} = U_k \square X_k \square Z, \text{ where } \tilde{X}_{k+1} = X_{k+p} \qquad (4)$$

with Z being a fuzzy relation $Z : \mathbf{U} \times \mathbf{X} \times \mathbf{X} \rightarrow [0, 1]$ given by the following membership function,

$$Z(u, x, y) = \max_{x_1, x_2, \ldots, x_{p-1} \in \mathbf{X}} [X_{k+1}(x_1)\ t\ X_{k+2}(x_2) \ldots t\ X_{k+p-1}(x_{p-1})\ t\ R(u, x, x_1, x_2, \ldots, x_{p-1}, y)], \qquad (5)$$

$$u \in \mathbf{U}, \quad x, y \in \mathbf{X}.$$

Proof. The thesis of the proposition may be easily proved by rewritting (2) as follows,

$$X_{k+p}(y) = \max_{u,x} \{U_k(u)\ t\ X_k(x)\ t \max_{x_1, x_2, \ldots} [X_{k+1}(x_1)\ t\ X_{k+2}(x_2)\ t \ldots t\ X_{k+p-1}(x_{p-1})\ t\ R(u, x, x_1, x_2, \ldots, x_{p-1}, y)]\} \qquad (6)$$

Insertion of (5) into (6) yields,

$$X_{k+p}(y) - \max_{u,x} [U_k(u)\ t\ X_k(x)\ t\ Z(u, x, y,)], \ y \in \mathbf{X} \qquad (7)$$

which completes the proof.

Bearing in mind the proposition stated above, we can easily reduce the order of the fuzzy model considering only fuzzy models of the first order without any loss of generality of analysis.

Now the following question arises: how the fuzzy model (1) can be constructed. Let us assume we have several fuzzy data representing the knowledge dealt with in the system being considered. Let us formalize it in the form of ordered triples,

$$U_1 \; X_1 \; X_2$$
$$U_2 \; X_2 \; X_3$$
$$\vdots$$
$$U_k \; X_k \; X_{k+1}$$

Now recall some results of theory of fuzzy relation equations with t-norms, [9] [11][12][14]. Denote by $R(U_k, X_k)$ and $U(X_k, R)$ families of fuzzy relations and fuzzy sets, respectively such that

$$R(U_k, X_k) = \{R: \mathbf{U} \times \mathbf{X} \times \mathbf{X} \rightarrow [0, 1] \mid U_k \,\square\, X_k \,\square\, R = X_{k+1}\} \qquad (8)$$

$$U(X_k, R) = \{U: \mathbf{X} \rightarrow [0, 1] \mid U \,\square\, X_k \,\square\, R = X_{k+1}\} \qquad (9)$$

Introduce moreover Ⓟ-compositions [9] of fuzzy sets and relation,

$$\hat{R}_k = (U_k \,\square\, X_k) \; Ⓟ \; X_{k+1} \qquad (10)$$

$$\hat{R}_k(u, x, y) = [U_k(u) \; t \; X_k(x)] \; \varphi \; X_{k+1}(y), \quad u \in \mathbf{U}, \; x, \; y \in \mathbf{X} \qquad (11)$$

$$\hat{U}_k = (X_k \,\square\, R) \; Ⓟ \; X_{k+1} \qquad (12)$$

$$\hat{U}_k(u) = \min_{y \in \mathbf{X}} \{\max_{x \in \mathbf{X}} [X_k(x) \; t \; R(u, x, y)] \; \varphi \; X_{k+1}(y)\} \qquad (13)$$

$$a \; \varphi \; b = \sup \{c \in [0, 1] \mid a \; t \; c \leq b\}, \; a, \; b \in [0, 1].$$

Then we have

Proposition 2.

(i) If $R(U_k, X_k) = \emptyset$ then $\hat{R}_k = \max R(U_k, X_k)$,

(ii) If $\bigcap_{k=1}^{K} R(U_k, X_k) = \emptyset$ then $\hat{R} = \bigcap_{i=1}^{K} \hat{R}_i$ belongs to every $R(U_k, X_k)$, $k = 1, 2, \ldots, K$.

The proof may be found in [14].

Moreover the following is valid,

Proposition 3.

If $U(X_k, R) = \emptyset$ then $\hat{U}_k = \max U(X_k, R)$

Thus the calculations of the fuzzy relations of the model are applied directly by taking the intersection of the fuzzy relations coming from respective triples of the fuzzy data. The order of the fuzzy equation and the choice of a proper t-norm form a topic of separate studies [10][12][13]. In general, the order of the model equation depends on the structure of linguistic information that is conveyed by a human being while describing the system in a linguistic fashion.

Let us consider now a problem of fuzzy control and construction of a fuzzy controller for a set of control rules.

FUZZY CONTROL AND FUZZY CONTROLLERS

As previously, let us pay our attention to the fuzzy model provided by (1). A goal of control is represented by a fuzzy set, $G : \mathbf{X} \rightarrow [0, 1]$. There is also a certain constraint imposed on fuzzy control and provided by a fuzzy set $C : \mathbf{U} \rightarrow [0, 1]$. Hence the problem of control is posed as follows,

-find a fuzzy control U_k that allows the fuzzy system (1) to achieve the goal G for a given state X_k.

Let us put down this formulation taking into account the model equation. Thus we have to solve the following equation

$$G = U_k \square X_k \square R \tag{14}$$

with respect to U_k. Denote

$$U(G, X_k) = \{U: \mathbf{U} \rightarrow [0, 1] \mid U \square X_k \square R = G\} \tag{15}$$

a family of fuzzy sets of control that allows the system to reach the goal from the specified fuzzy state X_k. We can immediately concludes, see Proposition 3, that for the nonempty family $U(G, X_k)$, the set $\hat{U}_k$ is the greatest element of it. In other words, if $U(G, X_k)$ is nonempty, we say the fuzzy model is (X_k, G) -controllable. It is worthwhile mentioning that the property of controllability is tied with the actual state and goal simultaneously; the system may be (X_k, G) -controllable and for $X_k' \neq X_k$ this property may be not preserved. Therefore the controllability is of a local, not of a global character. Let us discuss the case where $U(X_k, G)$ is empty. Then we may speak about approximate controllability of the fuzzy system. In order to measure this property it is worthwhile introducing a definition of equivalence of two fuzzy sets, say A and B, defined in the same space. A degree of equivalence of A and B, $[A = B]$, is calculated via formula,

$$[A = B] = \{\min_{x \in \mathbf{X}} [A(x) \varphi B(x)]\} \; t \; \{\min_{x \in \mathbf{X}} [B(x) \varphi A(x)]\} \qquad (16)$$

where "t" and φ are joint together. The background of this formula comes directly from multiplevalued logic, where equivalence or two propositions $A \leftrightarrow B$ is put down as meet of two implications $A \leftrightarrow B = (A \rightarrow B) \; \& \; (B \rightarrow A)$. Here " $\rightarrow$ " is modelled by the φ operator, while & is represented by the corresponding t-norm. Note that if A and B have different supports, $\text{supp}(A) \neq \text{supp}(B)$, $\text{supp}(A) = \{x \in \mathbf{X} \; A(x) > 0\}$, then the degree of equivalence (16) becomes zero. Imposition of a threshold level $\alpha \in [0, 1]$ allows us to consider the above degree of equivalence as a fuzzy set, $[A = B] : [0,1] \rightarrow [0,1]$. Now the definition (16) is read as [5]

$$[A = B](\alpha) = \min_{x \in \mathbf{X}} [A_\alpha(x) \varphi B_\alpha(x)] \; t \; \min_{x \in \mathbf{X}} [B_\alpha(x) \varphi A_\alpha(x)] \qquad (17)$$

where A_α, B_β are the original fuzzy sets A and B modified by α, $A_\alpha = A \cup \underline{\alpha}$, $B_\alpha = B \cup \underline{\alpha}$. $\underline{\alpha}$ is a fuzzy set $\underline{\alpha}(x) = \alpha$, for all $x \in \mathbf{X}$. The main idea that has been used in this construction is the following: small values of the membership function are meaningless and equivalence of fuzzy sets should be evaluated with regard to higher ones. In fact, if $A = B$ (equality of the membership functions of A and B), then $[A = B](\alpha) = 1$ for all α. On the other hand, if $[A = B](1) = 1$ and 0 otherwise, then A and B are completely different. Obviously $[\cdot]$ is a non-decreasing function of α. The fuzzy index of equivalence will be treated as the controllability index the use of which will be explained below,

-for the fuzzy model known, G and X_k provided, calculate the fuzzy control as the $\textcircled{\varphi}$-composition of $X_k \square R$ and G,

$$\hat{U}_k = (X_k \square R) \textcircled{\varphi} G \qquad (18)$$

-calculate the fuzzy set of state F caused by $\hat{U}_k$,

$$\hat{U}_k \square X_k \square R = F, \tag{19}$$

-compute [F = G] for all α's.

If [F = G] is identically equal to 1, we have absolute controllability, while the case

$$[F = G](\alpha) = \begin{cases} 1, & \alpha = 1 \\ 0, & \alpha \neq 1 \end{cases} \tag{20}$$

specifies total uncontrollability. Thus we are able to determine the fuzzy control in an efficient fashion. In situation where we consider the fuzzy constraint given by the fuzzy set C, the calculations of the fuzzy control are performed in a slightly modified manner.

Now the fuzzy control should be chosen in such a way that it should satisfy the goal G and the constraint C simultaneously. Define the fuzzy relation W formed as the cartesian product of G and C,

$$W = G \times C \tag{21}$$

$$W(y, u) = G(y) \ t \ C(u), \quad y \in \mathbf{Y}, \ u \in \mathbf{U}. \tag{22}$$

Then we require equality of two relations, where the first is viewed as cartesian product resulting from the fuzzy state X_{k+1} and C. This yields,

$$W = (U_k \square X_k \square R) \times C \tag{23}$$

or equivalently by the use of the membership functions,

$$W(y, v) = \max_{u,x} [U_k(u) \ t \ X_k(x) \ t \ R(u, x, y)] \ t \ C(v) = \max_{u,x} [U_k(u) \ t \ C(v) \ t \ X_k(x) \ t \ R(u, x, y)], \quad y \in \mathbf{X}, \ v \in \mathbf{U}.$$

Then the greatest fuzzy set of control $\hat{U}_k$ is obtained from the formula,

$$\hat{U}_k(u) = \min_{v,y} \{[C(v) \ t \ X_k(x) \ t \ R(u, x, y)] \ \varphi \ W(y, v)\} \tag{24}$$

The fuzzy controller specified by solving the equation of the fuzzy model can be helpful as an alternative for construction of the fuzzy controllers known in industrial applications. Such controllers consists of several control rules "if then" statements where antecedents are fuzzy states, while consequents are fuzzy actions control, if $X^{(i)}$ then $U^{(i)}$, $i = 1, 2, \ldots, N$. Let us generate fuzzy control for the fuzzy states as formed before. For the fuzzy model at hand, the fuzzy control may be immediately determined by straightforward calculations. Then we get a set of control rules

$$X^{(1)} \;---\; U^{(1)}$$
$$X^{(2)} \;---\; U^{(2)}$$

$$X^{(N)} \;---\; U^{(N)}$$

with the controllability indices associated with every control rule, that manifests feasibility of attaining the fuzzy goal from the actual state represented by $X^{(i)}$. Next, if it is more convenient for any further use, all of the rules given above are collected together in the form of fuzzy relation equation of the controller,

$$U_k = X_k \,\square\, T \tag{25}$$

with T fulfilling a family of equations,

$$U^{(i)} = X^{(i)} \,\square\, T, \; i = 1, 2, \ldots, N. \tag{26}$$

We are able to follow another path, assuming firstly the structure of the fuzzy controller as given by expression,

$$U_k = X_k \,\square\, T \tag{27}$$

and making use of the model equation.

Proposition 4.

For the fuzzy model described by (1) and the equation of the fuzzy controller specified by (27) the fuzzy relation T calculated with the aid of fuzzy sets of state $X^{(i)}$, $i = 1, 2, \ldots, N$ is equal to

$$T(z, u) = \min_i \min_y \{\max_x [X^{(i)}(z) \; t \; X^{(i)}(x) \; t \; R(u, x, y)] \; \varphi \; G(y)\} \tag{28}$$

$z \in \mathbf{X}$, $u \in \mathbf{U}$.

Proof. Let us insert the equation of the controller (27) into the model equation,

$$X_{k+1} = U_k \square X_k \square R = X_k \square T \square X_k \square R, \tag{29}$$

viz.

$$X_{k+1}(y) = \max_{u,x} \{\max_{z} [X_k(z) \; t \; T(z, u) \; t \; X_k(x) \; t \; R(u, x, y)]\}$$

$$= \max_{u,x,z} [X_k(z) \; t \; T(z, u) \; t \; X_k(x) \; t \; R(u, x, y)], \; y \in \mathbf{X}. \tag{30}$$

This for $X_k = X^{(i)}$, i = 1, 2, ..., N inverts into a set of equations,

$$G = X^{(1)} \square T \square X^{(1)} \square R$$

$$G = X^{(2)} \square T \square X^{(2)} \square R$$

$$\vdots \tag{31}$$

$$G = X^{(N)} \square T \square X^{(N)} \square R$$

Calculating, say $T^{(i)}$, for the i-th equation we get,

$$T^i = (X^i \square X^i \square R) \; \varphi \; G \tag{32}$$

Further on, T results as intersection of T^i's,

$$T = \bigcap_{i=1}^{N} T^i \tag{33}$$

Expressing the membership function of T completes the proof.

The controllability index is calculated with the aid of the index $[G = \tilde{X}^{(i)}]$ where $\tilde{X}^{(i)}$ denotes the fuzzy set obtained from the model equation for $X_k = X^{(i)}$. A global index of controllability is taken as minimum of the indices coming from the individual rules [5] $\min_i [G = \tilde{X}^{(i)}]$.

NUMERICAL STUDIES

In what follows we intend to illustrate the ways of deriving the fuzzy control by a numerical example. The data set comes from [3] and is often considered for purpose of testing new identification algorithms. This well known data set consists of 296 pairs of input-output observations that come from a gas furnace and they are sampled in a fixed interval of a length of 9 sec. The input is the gas flow rate into the furnace while the state is the concentration of CO_2[%] in the exhaust gases.

In [18] a fuzzy model has been proposed and tested. Next in [12] [13] we have proposed another model that ties input and state by a fuzzy relation. The accuracy of the results achieved is comparable with these produced by statistical models.

In our identification algorithm we deal not with single numerical values samples of the data set but with aggregates fuzzy sets defined in the input and state spaces, viz. **U** and **X**, respectively. Because we have no extra information about the structure of the data set, and both input and state are scalar not vector values, viz. one has some elements distributed along real line **R**, we apply a fuzzy clustering algorithm for generation membership functions of the linguistic labels for **U** and **X**. It may be done by e.g. Fuzzy Isodata (Fuzzy C-Means), a popular version of the fuzzy clustering algorithm cf. [2]. Speaking more formally it splits the data set into several classes, say c, and detects their centroids, that are also called fuzzy means. All the information about the membership functions is conveyed by a partition matrix $P = [p_{ik}]$, $i = 1, 2, \ldots, c$, $k = 1, 2, \ldots, n$ where "n" stands for the number of objects being clustered. Thus the membership function defined in a finite number of elements data of the j-th cluster is provided by the j-th row of the partition matrix, $[p_{j1}\ p_{j2} \ldots p_{jn}$, $j = 1, 2, \ldots, c]$. Hence we are able to generate the fuzzy partition of the universe of **U** and **X** that reflects the structure of the data at our disposal. We cannot neglect the problem of choice of the number of clusters "c" used to split the data set; it may vary in a wide range between 2 and n. It is emphasized that their number should be proposed externally by the user who intends to use the fuzzy model. One caution is reasonable to formulate: if the number of clusters increases then a highter level of precision is obtained, but we may face the fact that so many clusters may not have any linguistic relevance. Simply, it is an example of the incompatibility principle [22]. Thus if the user intends to put down the following linguistic labels in the space **U**: high gas flow rate, medium, zero, negative ones, we may insert for the clustering method the parameter "c" equal to 4. In further studies we fix the number of the fuzzy labels equal to 5. As shown in [13] this number of linguistic categories gives enough good concordance of the data set with the model constructed. The Fuzzy Isodata generated the clusters whose functional form may be approximated by bell-shape function $\exp[-(x - m_i)^2/\sigma^2_i]$ where m_i denotes the fuzzy means, while σ_i is responsible for the spread. We have the following membership functions,

the control space **U**		the state space **X**	
m_i	σ_i	m_i	σ_i
-1.93	4.68	48.03	4.50
-0.90	3.84	50.97	3.13
-0.01	0.77	53.47	2.54
0.83	0.73	56.00	2.50
1.81	0.89	58.55	2.00

The reference fuzzy sets defined in X are also displayed in Fig. 1.

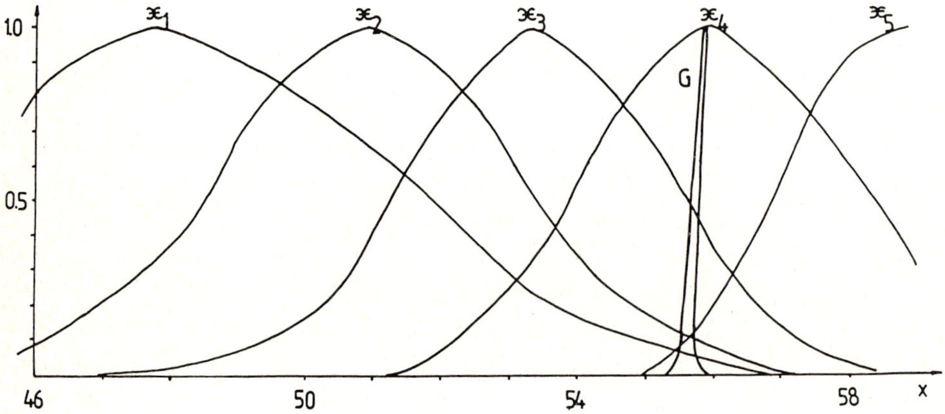

Fig. 1.

Note a certain overlap between the fuzzy labels constructed. Maybe the shape of the membership functions of the limit fuzzy sets (viz. $\mathcal{X}_1$, $\mathcal{X}_5$, $\mathcal{U}_1$, $\mathcal{U}_5$) is somewhat suprising, perhaps one should expect the values of the membership functions to be equal to 1 for all the values lying on left (right) of their centers. Nevertheless it is evident that these ranges of values have no reflection in the data set and such an extrapolation would be senseless. The union of all the reference fuzzy sets "cover" the entire space **U** or **X**, viz.

$$\forall x \in \mathbf{X}, \ \max_i \mathcal{X}_i(x) > 0, \quad \forall u \in \mathbf{U}, \ \max_i \mathcal{U}_i(u) > 0, \qquad (34)$$

where $\mathbf{X} = [\min_{1\leq j\leq 296} x_j, \max x_j]$, $\mathbf{U} = [\min_{1\leq j\leq 296} u_j, \max u_j]$.

Then any fuzzy set is immediately expressed by these reference fuzzy sets. Let X

: $\mathbf{X} \rightarrow [0, 1]$ be any fuzzy set in $\mathbf{X}$. It is transformed into a fuzzy set : $\mathcal{X} \rightarrow [0, 1]$, i = 1, 2, ..., 5 by taking possibility measure of X with respect to $\mathcal{X}_i$[23]

$$\text{Poss}\,(X \mid \mathcal{X}_i) = \sup_{x \in \mathbf{X}} [\min(X(x),\ \mathcal{X}_i(x))] \tag{35}$$

For X being a nonfuzzy quantity $X(x) = \delta\,(x - x_0) = \begin{cases} 1, & x = x_0 \\ 0, & x \neq x_0 \end{cases}$ the above formula is read as

$$\text{Poss}\,(X \mid \mathcal{X}_i) = \mathcal{X}_i(x_0) \tag{36}$$

Conversely, any fuzzy set $\mathcal{X}$ could be mapped in the space $\mathbf{X}$. For instance having $\mathcal{X} = [a_1, a_2, \ldots, a_5]$ we read it as: X matches $\mathcal{X}_i$ with a grade possibility measure equal to a_i, i = 1, 2, ..., 5.

Numerical studies and some optimization procedures lead us to the fuzzy model described by the fuzzy relation equation of the first order with delay with max-product composition,

$$X_{k+1} = U_{k-2} \,\square\, X_k \,\square\, R \tag{37}$$

$U_{k-2} \in F(\mathcal{U})$, X_k, $X_{k+1} \in F(\mathcal{X})$. It is remarkable to underline that this fuzzy model works with relationships given by the reference fuzzy sets, viz. the fuzzy relation that embraces ties between the fuzzy labels. It is represented in the matrix shown in Table 1. The most significant elements of this table are underlined. The entries of this matrix are viewed as possibilities assigned to the sentences evaluating the system behaviour, and having a format,

-if the state in the k-th time moment and the control in the (k - 2)th time moment then the state in the (k + 1)th moment,

with the possibility assigned to it.

For instance we have,

$$\begin{matrix} k-2 & k & & k+1 \\ (\,\mathcal{U}_2 & \mathcal{X}_4\,) & \rightarrow & \mathcal{X}_3 \end{matrix}$$

with the value of the possibility measure equal to 0.41. Now let us design the fuzzy control algorithm following the method introduced in the previous section. One has to pay attention that we deal with the model with delay in control, which requires a certain modification. Performing iterations one has,

Table 1. The fuzzy relation of the fuzzy model

$\mathcal{X}$ (rows) × $\mathcal{X}$ (columns), $\mathcal{U}_1$:

0.03	0.01	0.01	0.01	0.01
0.02	0.04	0.01	0.05	0.02
0.01	0.03	0.39	0.92	0.05
0.00	0.02	0.05	0.76	0.74
0.01	0.02	0.04	0.11	0.98

$\mathcal{U}_2$:

0.04	0.03	0.03	0.01	0.01
0.04	0.64	0.73	0.04	0.02
0.01	0.05	0.83	0.72	0.04
0.01	0.02	0.41	0.92	0.35
0.01	0.02	0.04	0.50	0.95

$\mathcal{U}_3$:

0.48	0.49	0.05	0.02	0.01
0.13	0.89	0.39	0.03	0.01
0.02	0.29	1.00	0.13	0.01
0.01	0.04	0.49	0.98	0.18
0.01	0.01	0.04	0.66	0.99

$\mathcal{U}_4$:

0.79	0.32	0.04	0.02	0.01
0.23	0.96	0.21	0.02	0.00
0.04	0.54	0.81	0.60	0.04
0.02	0.05	0.64	0.69	0.05
0.00	0.01	0.05	0.03	0.13

$\mathcal{U}_5$:

0.97	0.09	0.04	0.02	0.01
0.51	0.75	0.06	0.02	0.01
0.05	0.27	0.03	0.01	0.00
0.03	0.02	0.03	0.02	0.02
0.02	0.01	0.00	0.01	0.02

$$X_{k+1} = U_{k-2} \square X_k \square R$$

$$X_{k+2} = U_{k-1} \square X_{k+1} \square R \tag{38}$$

$$X_{k+3} = U_k \square X_{k+2} \square R$$

Insertion of both of the first formulas into the third one yields

$$X_{k+3} = U_k \square U_{k-1} \square U_{k-2} \square X_k \square R \square R \square R. \tag{39}$$

Then if the goal is given by the fuzzy set $G \in F(\mathcal{X})$, we force equality (39) for $X_{k+3} = G$. Now the fuzzy control we are interested in depends on G and the control actions U_{k-2}, U_{k-1} used before. The greatest fuzzy set of control $\hat{U}_k$ is calculated according to

$$\hat{U}_k = (U_{k-1} \square U_{k-2} \square X_k \square R \square R \square R) \textcircled{\varphi} G \tag{40}$$

The fuzzy goal is specified by the membership function G = [0.059 0.11 0.49 0.97 0.11] in the space of the reference fuzzy sets of state (see Fig. 1) and the fuzzy states $\mathcal{X}_1\ \mathcal{X}_2\ \mathcal{X}_3\ \mathcal{X}_4\ \mathcal{X}_5$. We have the following control rules. They depend on the fuzzy controls for the (k - 2), (k - 1)th time moments. Table 2. displays the control rules calculated.

Table 2. Rules of the fuzzy control derived from the fuzzy model

$\mathcal{U}_1, \mathcal{U}_1$:	$\mathcal{X}_1 \rightarrow$	[0.32	0.33	0.32	1.00	1.00]
	$\mathcal{X}_2 \rightarrow$	[0.20	0.21	0.20	1.00	1.00]
	$\mathcal{X}_3 \rightarrow$	[0.16	0.17	0.16	1.00	1.00]
	$\mathcal{X}_4 \rightarrow$	[0.16	0.16	0.13	1.00	1.00]
	$\mathcal{X}_5 \rightarrow$	[0.12	0.12	0.17	0.88	1.00]
$\mathcal{U}_2, \mathcal{U}_2$:	$\mathcal{X}_1 \rightarrow$	[0.35	0.51	0.37	0.34	0.34]
	$\mathcal{X}_2 \rightarrow$	[0.28	0.42	0.30	0.28	0.28]
	$\mathcal{X}_3 \rightarrow$	[0.22	0.46	0.38	0.29	0.35]
	$\mathcal{X}_4 \rightarrow$	[0.18	0.18	0.17	0.39	0.64]
	$\mathcal{X}_5 \rightarrow$	[0.12	0.13	0.12	0.60	0.97]
$\mathcal{U}_3, \mathcal{U}_3$:	$\mathcal{X}_1 \rightarrow$	[0.60	0.27	0.19	0.17	0.18]
	$\mathcal{X}_2 \rightarrow$	[0.35	0.22	0.16	0.15	0.15]
	$\mathcal{X}_3 \rightarrow$	[0.20	0.27	0.19	0.18	0.18]
	$\mathcal{X}_4 \rightarrow$	[0.16	0.17	0.16	0.27	0.33]
	$\mathcal{X}_5 \rightarrow$	[0.11	0.12	0.11	0.41	0.50]
$\mathcal{U}_4, \mathcal{U}_4$:	$\mathcal{X}_1 \rightarrow$	[0.60	0.22	0.16	0.20	0.10]
	$\mathcal{X}_2 \rightarrow$	[0.38	0.19	0.13	0.12	0.12]
	$\mathcal{X}_3 \rightarrow$	[0.31	0.23	0.17	0.15	0.16]
	$\mathcal{X}_4 \rightarrow$	[0.31	0.42	0.31	0.28	0.29]
	$\mathcal{X}_5 \rightarrow$	[0.39	0.62	0.45	0.42	0.42]
$\mathcal{U}_5, \mathcal{U}_5$:	$\mathcal{X}_1 \rightarrow$	[1.00	0.37	0.13	0.08	0.06]
	$\mathcal{X}_2 \rightarrow$	[1.00	0.31	0.16	0.09	0.07]
	$\mathcal{X}_3 \rightarrow$	[1.00	0.38	0.27	0.19	0.15]
	$\mathcal{X}_4 \rightarrow$	[1.00	0.69	0.50	0.34	0.28]
	$\mathcal{X}_5 \rightarrow$	[1.00	1.00	0.76	0.52	0.42]

For every output of the fuzzy controller resulting from the model equation solved for the actual state and past controls known, the index of controllability $\xi(\alpha) = [G = \tilde{X}](\alpha)$, where X comes from the fuzzy model is calculated. Fig. 2 illustrates the index of controllability for the fuzzy control for various states in the k-th time moment and the past controls fixed and being equal to $\mathcal{U}_1$, $\mathcal{U}_1$.

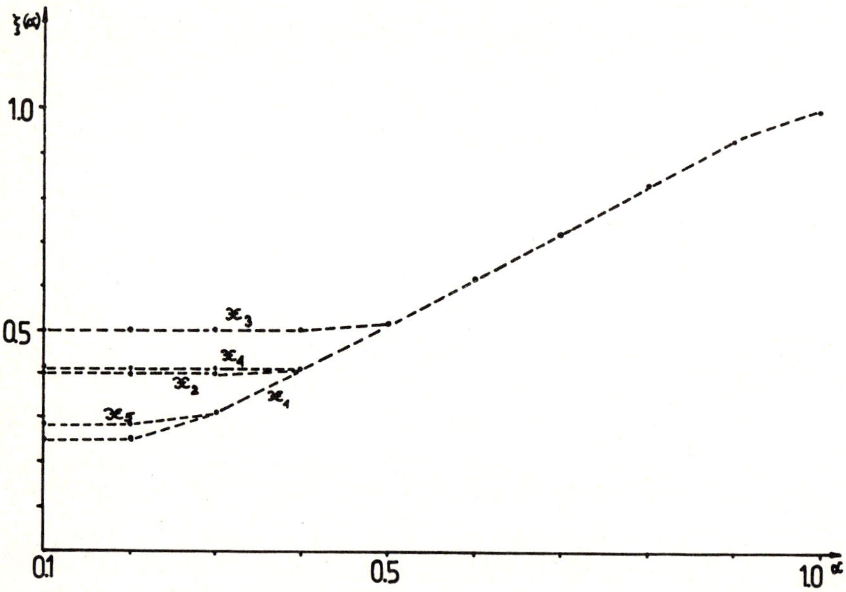

Fig. 2.

Among all the reference fuzzy sets of state the controllability index attains higher values for $\mathcal{X}_3$, for $\alpha = 0.1$ being equal to 0.5. This is not suprising, bearing in mind the fact the fuzzy goal bears a strong resemblance with $\mathcal{X}_3$. The lowest value is obtained for $\mathcal{X}_1$, $\xi(0.1) = 0.25$. This fact is in accordance with degree of matching of $\mathcal{X}_1$ and G, Poss($\mathcal{X}_i \mid G$). Summarizing these results we get,

Poss($\mathcal{X}_i \mid G$)	0.02	0.07	0.11	0.15	0.09
[G = $\mathcal{X}_i$](0.1)	0.254	0.403	0.497	0.411	0.290

The control rules that we have already derived will be put into the equation of the controller, that gives explicit formula for calculating the fuzzy control U for the fuzzy state known and the past controls,

$$U = X \square U_{k-2} \square U_{k-1} \square G \tag{41}$$

The fuzzy control depends on the state of the model, and the above equation is similar, but in a shape only, to the formula for feedback control in optimal control of deterministic or stochastic systems cf. [6]. Determination of the relation of the controller is performed with regard to the fuzzy sets specified before. The results of computations are collected in Table 3.

Table 3. The fuzzy relation of the controller

$$\mathcal{X} \downarrow \; \overrightarrow{\mathcal{U}} \quad \begin{bmatrix} 0.25 & 0.26 & 0.25 & 1.00 & 1.00 \\ 0.20 & 0.21 & 0.20 & 1.00 & 1.00 \\ 0.17 & 0.17 & 0.17 & 1.00 & 1.00 \\ 0.15 & 0.15 & 0.13 & 1.00 & 1.00 \\ 0.12 & 0.12 & 0.12 & 0.88 & 1.00 \end{bmatrix} \mathcal{U}_1, \mathcal{U}_1 \qquad \begin{bmatrix} 0.34 & 0.51 & 0.37 & 0.34 & 0.34 \\ 0.28 & 0.41 & 0.30 & 0.28 & 0.28 \\ 0.22 & 0.24 & 0.23 & 0.30 & 0.35 \\ 0.16 & 0.16 & 0.15 & 0.39 & 0.46 \\ 0.12 & 0.13 & 0.12 & 0.55 & 0.90 \end{bmatrix} \mathcal{U}_2, \mathcal{U}_2$$

$$\begin{bmatrix} 0.43 & 0.26 & 0.19 & 0.17 & 0.18 \\ 0.25 & 0.22 & 0.16 & 0.14 & 0.15 \\ 0.20 & 0.22 & 0.19 & 0.18 & 0.18 \\ 0.14 & 0.15 & 0.14 & 0.16 & 0.23 \\ 0.11 & 0.12 & 0.11 & 0.22 & 0.38 \end{bmatrix} \mathcal{U}_3, \mathcal{U}_3 \qquad \begin{bmatrix} 0.46 & 0.22 & 0.16 & 0.20 & 0.10 \\ 0.38 & 0.19 & 0.13 & 0.12 & 0.12 \\ 0.31 & 0.23 & 0.17 & 0.15 & 0.15 \\ 0.31 & 0.30 & 0.22 & 0.20 & 0.20 \\ 0.39 & 0.59 & 0.43 & 0.39 & 0.40 \end{bmatrix} \mathcal{U}_4, \mathcal{U}_4$$

$$\begin{bmatrix} 1.00 & 0.37 & 0.13 & 0.08 & 0.06 \\ 1.00 & 0.31 & 0.16 & 0.09 & 0.07 \\ 1.00 & 0.38 & 0.20 & 0.11 & 0.09 \\ 1.00 & 0.50 & 0.36 & 0.20 & 0.16 \\ 1.00 & 0.97 & 0.70 & 0.48 & 0.39 \end{bmatrix} \mathcal{U}_5, \mathcal{U}_5$$

The control calculated has to be interpreted as follows. Let U_k be the fuzzy set of control calculated via (41). It suggests which control among the linguistic labels $\mathcal{U}_1$, $\mathcal{U}_2$, $\mathcal{U}_3$, $\mathcal{U}_4$, $\mathcal{U}_5$ should be applied. Nevertheless we do not specify a certain nonfuzzy value that may be treated as the control action. There is no unique way of establishing the nonfuzzy value for the fuzzy set attached. N.B. the same problem holds in probabilistic decision-making; one can take median, averaged mean or modal value of a probability function expressed in a finite decision space. Here we follow the method of averaged mean. The fuzzy set in **U**, $[a_1\ a_2\ a_3\ a_4\ a_5]$ is represented by a real number $\bar{a} \in \mathbf{R}$ calculated as,

$$\bar{a} = \sum_{i=1}^{5} a_i \bar{x}_i \Big/ \sum_{i=1}^{5} a_i \tag{42}$$

where $\bar{x}_i$ stands for the centroid of the reference fuzzy set $\mathcal{U}_i$. The nonfuzzy characteristics obtained for the fuzzy controller are shown in Fig. 3.

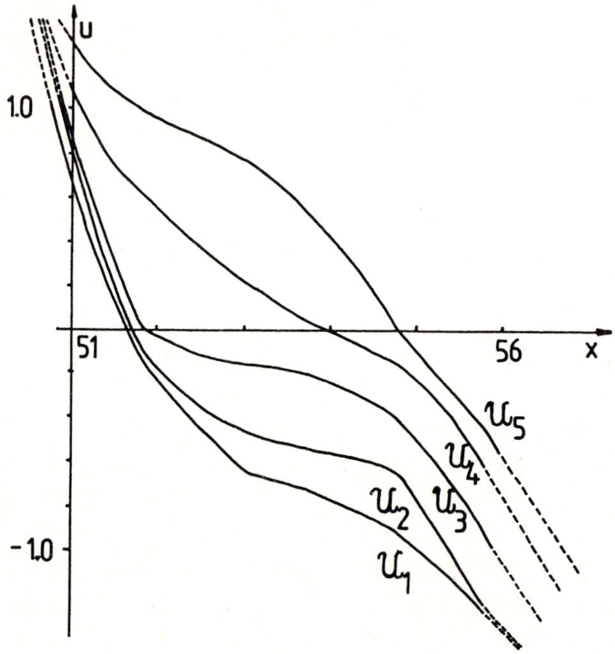

Fig. 3.

It is worth emphasizing the nonlinearity of the characteristics of the fuzzy controller. In a medium range of the values of the state variable the values of control do not depend heavily on x, -the characteristics are flat. For the smaller and greater values of the state variable slight changes of x generate significant changes of the control variable. This type of nonlinearity is in accordance with the general control strategy: the changes of the control variable should be small near the goal, and large when the actual state is far from the goal.

CONCLUDING REMARKS

We have discussed some problems of fuzzy control using a fuzzy model of a non-deterministic system. Several basic control policies have been considered and the construction of the fuzzy controller have been presented exhaustively. In general, the generated control rules may be viewed separately or may be combined in the form

of the fuzzy relation. This first method is more plausible when we are interested in analysis of individual rules (e.g. for comparative studies where some heuristic control rules are known). The second one is more convenient for on-line calculation of the fuzzy control. We have paid special attention to the controllability of the fuzzy system and ways of measuring this property. This approach to fuzzy control with the aid of fuzzy models is flexible and effective in the sense that we are able to perform deeper studies into the properties of fuzzy controllers than was possible when analyzing linguistic control protocols. Moreover, fundamental concepts present in modern deterministic or stochastic control theory may be investigated in the framework of the fuzzy models studied here. The importance of the fuzzy control algorithms given here for industrial practice becomes transparent in situations we are able to construct a linguistic model of the system being analyzed. In the paper we have used numerical data for illustration purposes; nevertheless one can easily construct a model for any linguistic data available. The representation of nonfuzzy and fuzzy quantities by means of fuzzy discretization makes the methodology of model derivation gereral enough. The linguistic statements creating the model are not general due to their descriptive power. They may be reasonable as computer assisted means for training new process control staff because they recall some linguistic terms of control resulting from control protocols of selected operators of the process. Therefore it is of interest to study a two-stage procedure for model construction and control that consists of two levels:
-first level at which we play with fuzzy sets and derive general control policy,
-second evel which makes use of the knowledge of the first level for deriving "local" nonfuzzy models and proposing "local" control algorithms.

REFERENCES

[1] J. F. Baldwin, B. W. Pilsworth: A model of fuzzy reasoning through multi-valued logic and set theory, Int. J. Man-Machine Studies, 11, 351-380, 1979

[2] J. C. Bezdek: Pattern Recognition with Fuzzy Objective Function Algorithms, Plenum Press, New York, 1981

[3] G. E. Box, G. M. Jenkins: Time Series Analysis: Forecasting and Control, Holden Day, San Francisco, 1970

[4] M. Braae, D. A. Rutherford: Theoretical and linguistic aspects of the fuzzy logic controller, Automatica, 15, 535-577, 1979

[5] S. Gottwald, W. Pedrycz: Solvability of fuzzy relational equations and manipulation of fuzzy data, Fuzzy Sets and Systems, submitted.

[6] H. Kwakernaak, R. Sivan: Linear Optimal Control Systems, J. Wiley, New York, 1972

[7] E. H. Mamdani, J. J. Ostergaard, E. Lembessis, Use of fuzzy logic for implementing rule-based control of industrial processes, in: Fuzzy Sets and Decision Analysis (H. J. Zimmermann, L. A. Zadeh, B. R. Gaines eds.), North Holland, New York, 429-445, 1984

[8] M. Mizumoto, H. J. Zimmermann: Comparison of fuzzy reasoning methods, Fuzzy Sets and Systems, 8, 253-283, 1982

[9] W. Pedrycz: Fuzzy relational equations with triangular norms and their resolution, Busefal, 11, 24-32, 1982

[10] W. Pedrycz: Numerical and applicational aspects of fuzzy relational equations, Fuzzy Sets and Systems, 11, 1-18, 1983

[11] W. Pedrycz: Fuzzy relational equations with generalized connectives and their applications, Fuzzy Sets and Systems, 10, 185-201, 1983

[12] W. Pedrycz: Construction of fuzzy relational models. In: Proc. Cybernetics and Systems Research, Vienna (R. Trappl, ed.), North Holland, Amsterdam, 545-549, 1984

[13] W. Pedrycz: An identification algorithm in fuzzy relational systems, Fuzzy Sets and Systems, 13, 153-167, 1984

[14] W. Pedrycz: Applications of fuzzy relational for methods of reasoning in presence of fuzzy data, Fuzzy Sets and Systems, submitted.

[15] K. E. Peray, J. J. Wadell: The Rotary Cement Kiln, Chemical, New York, 1972

[16] B. Schweizer, A. Sklar: Probabilistic Metric Spaces, North Holland, Amsterdam, 1983

[17] T. Takagi, M. Sugeno: Derivation of fuzzy control rules from human operator's control actions, Proc. Symp. Fuzzy Information, Knowledge Representation and Decision Analysis, Marseille, 55-60, July 1983

[18] R. M. Tong: The evaluation of fuzzy models derived from experimental data, Fuzzy Sets and Systems, 4, 1-12, 1980

[19] Y. Tsukamoto: An approach to fuzzy reasoning method, In: Advances in Fuzzy Set Theory and Applications (M. M. Gupta ed.), North Holland, Amsterdam, 407-428, 1975

[20] D. Willaeys: The use of fuzzy sets for the treatment of fuzzy information, Fuzzy Sets and Systems, 3, 1981

[21] L. A. Zadeh: Toward a theory of fuzzy systems, In: Aspects on Network and System Theory (R. E. Kalman, N. De Claris eds.), Holt, Rinehart, Winston, New York, 209-245, 1971

[22] L. A. Zadeh: Outline of a new approach to the analysis of complex systems and decision processes, IEEE Trans. Systems, Man and Cybernetics, 1, 28-44, 1973

[23] L. A. Zadeh: Fuzzy sets as a basis for a theory of possibility, Fuzzy Sets and Systems, 1, 3-28, 1978

Industrial Applications of Fuzzy Control
M. Sugeno (ed.)

Human Operator's Fuzzy Model in Man-Machine System with a Nonlinear Controlled Object

Kozo Matsushima* Hideyuki Sugiyama**

* Institute of Engineering Mechanics, University of Tsukuba
Sakura, Niihari, Ibaraki, JAPAN
** Graduate School of Engineering, University of Tsukuba
Sakura, Niihari, Ibaraki, JAPAN

Several approaches have been proposed in order to analyze the human operator's behavior in manual control systems. However, there are few research works concerning the analysis of the human's control behavior for a controlled object with nonlinearity such as saturation, dead band and so on. We propose a new method, by using a fuzzy model, for the analysis of the human's control behavior in the above-mentioned systems. We investigate an operator's pursuit tracking behavior for a step input and construct its fuzzy model consisting of linguistic control rules. These rules are described in terms of the control deviation and the change rate of the controlled variable on the assumption that the operator's inference is carried out discretely by responding to the states of the system. As the result, the fuzzy model is found useful for simulating and analyzing the human operator's control behavior in manual control systems.

I Introduction

Many research works concerning with man-machine systems have been performed. These works mainly study the human operator's control behavior for linear controlled objects by means of transfer function methods or state variable methods [1],[2].

The authors have analyzed the human's control behavior for a controlled object with dead band by a model consisting of two or more transfer functions which are successively selected by responding to the state of the system[3].

But, these approaches are not always sufficient to make the human's control strategy clear, especially so in the case of the controlled object with some kind of nonlinearity. The purpose of this study is to analyze the human's control behavior in a man-machine system by the aid of fuzzy sets[4] and to make the human's control strategy more clear.

A servomechanism with torque saturation is used as a controlled object controlled by the human operator. The experiments set up are the pursuit tracking task by visual feedback to a step input.

Using the experimental data, a fuzzy model of the human operator consisting of linguistic control rules is constructed. These rules are described in terms of the control deviation and the change rate of the controlled variable on the assumption that the operator's inference is carried out discretely by responding to the state of the system.

As the result, it is shown that the fuzzy model proposed in this study is found useful for simulating and analyzing the human operator's control behavior in manual control systems.

II Experiments for obtaining Human Operator Responses

The experiments are performed by the experimental set-up shown in Fig. 1. The controlled variable is displayed with the system input on the CRT.

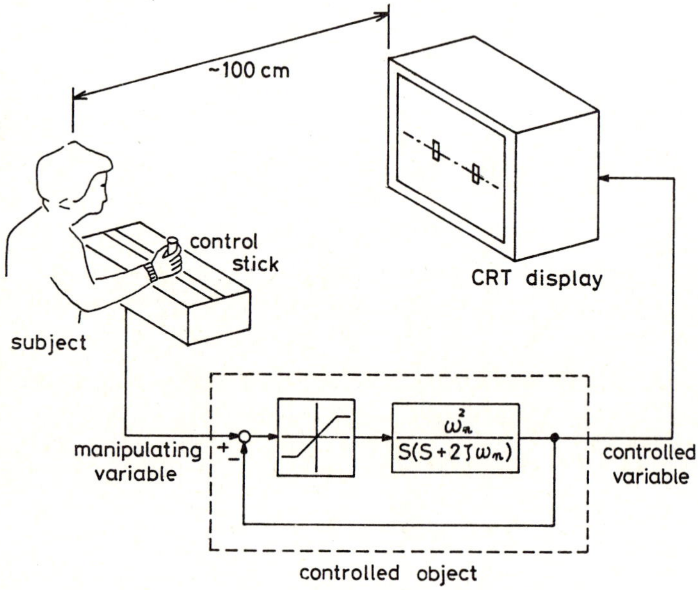

Fig.1 System layout

The controlled object is simulated by a micro computer and the dynamics of its linear part is described by eqs.(1) and (2)

$$\begin{bmatrix} x_1(k) \\ x_2(k) \end{bmatrix} = \begin{bmatrix} 1 & 0 \\ 0 & \alpha \end{bmatrix} \begin{bmatrix} x_1(k-1) \\ x_2(k-1) \end{bmatrix} + \begin{bmatrix} \Delta t \\ \alpha \cdot \Delta t \end{bmatrix} w(k) \tag{1}$$

$$c(k) = \left[\frac{\omega_n}{2\zeta}, -\frac{\omega_n}{2\zeta}\right] \begin{bmatrix} x_1(k) \\ x_2(k) \end{bmatrix} \tag{2}$$

where $w(k)$ is the output of saturation element, $c(k)$ is the controlled variable, Δt is the sampling period and $\alpha = \exp(-2\zeta\omega_n\Delta t)$. Numerical values of parameters ω_n and ζ are 5.64 rad/sec and 0.47 respectively to take account of dynamics of the remote mini manipulator which has been developed in the author's laboratory[5]. The magnitude of the system input is 13 cm on CRT screen. Subjects are required to manipulate a joystick for pursuit tracking the controlled variable to the system input on the CRT.

Table 1 Experimental factors

factor	level
saturation	L1 51.99 cm L2 6.50 cm L3 2.60 cm L4 1.29 cm
subject	S1,S2,S3,S4

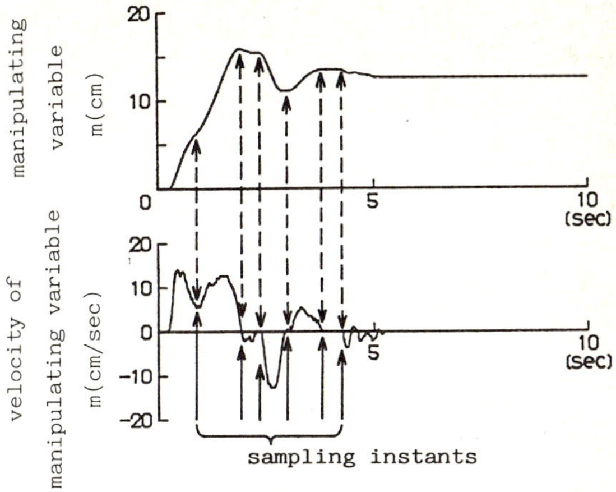

Fig.2 Sampling instant to decide the successive control action

The factors and levels of the experiments are shown in Table 1. In the table, the height of saturation is converted into the length on CRT.

Total number of trials of each subject are 40 trials, that is, 10 trials for each level of saturation. Through the observation of all experimental data, four subjects may be classified into two groups. One of them, (S1,S2), called Group 1, is one whose behaviors seem to be bang-bang.

The control behavior of another one, (S3,S4), called Group 2, is very moderate and almost all of them show responses without overshoot to step input.

III Fuzzy Model of Human Operator

In the experiments of this study, subjects can observe only two state variables, i.e., the control deviation, e(cm), and the change rate of the controlled variable, cc(cm/sec). Therefore it is assumed that each subject decides the value of the manipulating variable through observing these state variables.

Fig. 2 shows an example of a wave form of the manipulating variable (movement of joystick).

By the inspection of the data of the manipulating variable as exemplified in Fig. 2, it is presumed that the subject discretely decides his successive control actions by sampling the state of the system and generates tentative commands by which his neuro-muscular system is excited, because operators may not make the magnitude of the control deviation on CRT correspond exactly with the magnitude of the movement of the joystick. The sampling instants may be observed as the time instants when the manipulating variable takes a paticular value, e.g., its extreme or inflection one. And its sampling period depends on the states e and cc. The values of e and cc at each sampling instant obtained from experimental data are

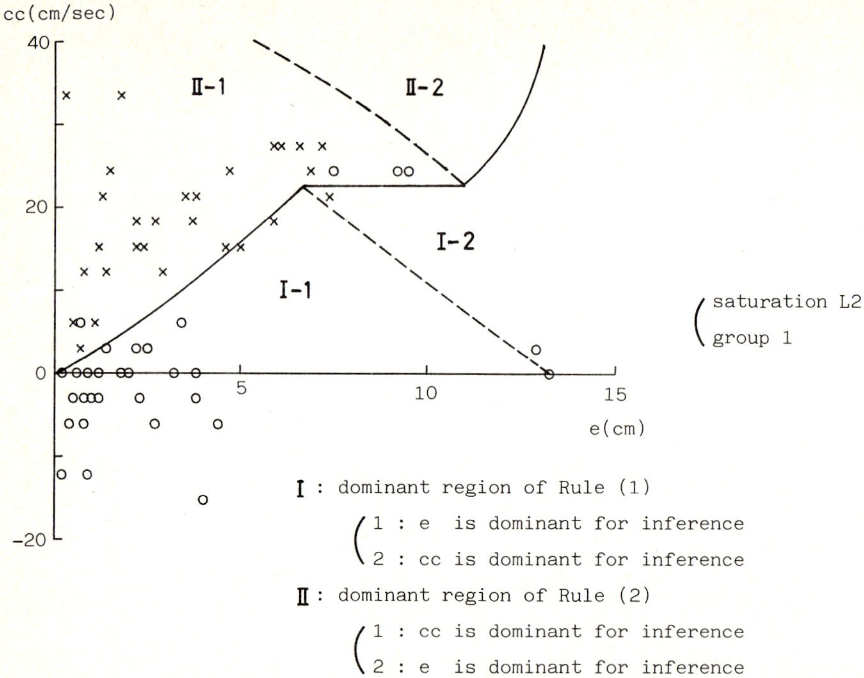

Fig.3 Sampling points plotted on e-cc plane and boundary lines

plotted for each factor as shown in Fig. 3 (in Fig. 3, the meaning of bold line and dashed line will be explained later.) In Fig. 3, the abscissa shows the absolute value of control deviation e and the ordinate the change rate of the controlled variable cc. The sign of cc is taken as positive when the controlled variable moves toward the system input and vice-versa. Plots "O" show the points where the manipulating variable is increased and plots "X" show the points where the manipulating variable is decreased.

III-1 Preliminary considerations concerning linguistic control rules

In order to construct a fuzzy model of the human operator, at first, some reasonable control rules are selected from possible control ones obtained through observing the experimental data. Then it is examined how these control rules contribute to the control action.

By plotting all data on the e-cc plane as in Fig. 3, the following operator's common actions are found out.

(1) if e is big and cc is small, then the manipulating variable is increased.

(2) if e is small and cc is big , then the manipulating variable is decreased,

and in the region cc $\leq$ 0, almost all control actions are performed to increase the manipulating variable. The sampling period used in order to decide the successive control action is

(1) long when e is big and

(2) short when cc is big

Under the above considerations, a fuzzy model to simulate the operator's control behavior is constructed by a micro computer as shown in Fig. 4.

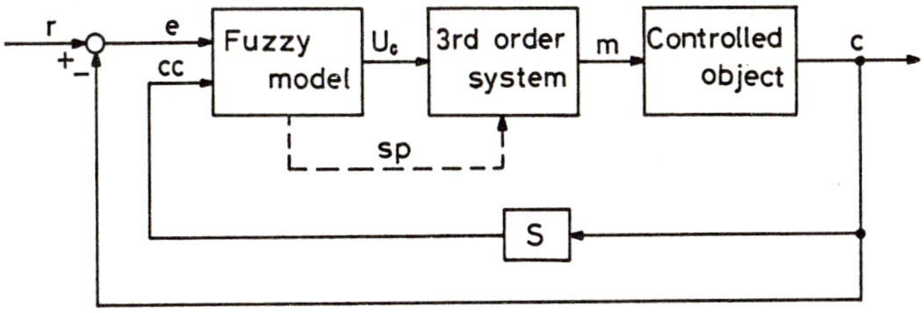

Fig.4 Block diagram of fuzzy control model

The fuzzy model outputs a step-like tentative command Uc which is generated by relating to e and cc and is held constant during the sampling period. This process may simulate the input-output relation of operator's central nervous system. A 3rd order system which outputs the real manipulating variable simulates the neuro-muscular system as a signal smoother. The fuzzy model consists of the fuzzy control rules which describe the operator's control strategy and the algorithm in order to get resultant numerical values by fuzzy inference.

The fuzzy control rules used here consist of two sets. One of them is that which infers the sampling period sp as

(a) if $|e| = B$ then $sp = B$,

(b) if $|cc| = B$ then $sp = S$.

The other one is a set of rules used to generate the increment of the tentative command, called u_c, as the input of the neuro-muscular system as follows,

(1) if e = PB and cc = PS then u_c = PB,

(2) if e = PS and cc = PB then u_c = NB,

(3) if e = PS and cc = NB then u_c = PB,

(4) if e = PS and cc = NS then u_c = PS,

and where B, S, PB, PS, NB and NS represent big, small, positive big, positive small, negative big and negative small respectively. The sign of u_c is taken to be positive in the direction which corresponds to the joystick being moved toward the right by the operator.

If e is negative then the above rules are changed as e = PB to e = NB, e = PS to e = NS, u_c = PB to u_c = NB, u_c = NB to u_c = PB and u_c = PS to u_c = NS.

Membership functions of fuzzy sets labeled as B and S are defined as follows, for the control rules (1) ∿ (4)

$$\mu_B(x) = (x/x_o)^h, \qquad \mu_S(x) = (1-x/x_o)^{h'} \qquad x \in [0,x_o] \qquad (3)$$

and for the control rules (a) and (b)

$$\mu_S(sp) = \{1-(sb-sp)/(sb-ss)\}^{h''}$$
$$\mu_B(sp) = \{(sp-ss)/(sb-ss)\}^{h'''} \qquad sp \in [ss,sb] \qquad (4)$$

The algorithm for obtaining the numerical value of u_c from the above-mentioned control rules is from Sugeno's method[6] as shown in Fig. 5.

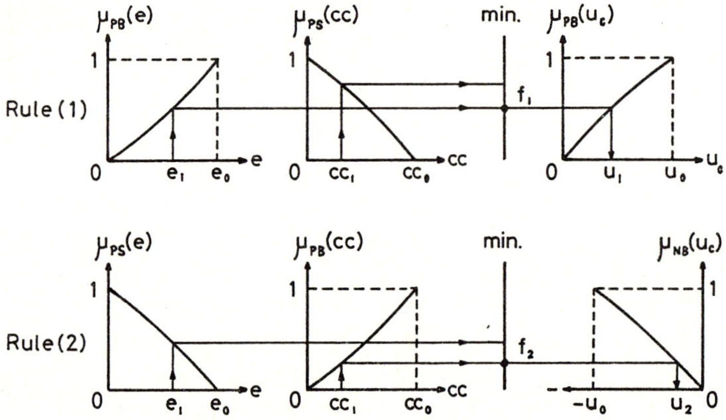

Fig. 5 Algorithm of fuzzy model

Fig. 5 shows an example to explain the way to get u_c from the control rules (1) and (2), where u_c is the increment in the tentative command and

$$u_c = \frac{f_1 u_1 + f_2 u_2}{f_1 + f_2} \qquad (5)$$

$$f_1 = \min(\mu_{PB}(e), \mu_{PS}(cc)), \qquad f_2 = \min(\mu_{PS}(e), \mu_{PB}(cc)). \qquad (6)$$

Changing the point of view, eq.(5) is interpreted as follows, the weights f_1 and f_2 are understood as the grade which each control rule contributes to the operator's inference in his control strategy and, concerning each control rule, one of either the state variables e or cc which decides f's value directly relates to the inference used to generate u_c.

Therefore, if it is possible to simulate the operator's control behavior by using the above-mentioned sets of control rules and the relating algorithm, then it may be infered which control rule is mainly weighted and which state variable is noticed by the operator to decide his control action.

Simulation of the operator's control behavior is executed by adjustment of the support of each fuzzy set, the hedge of each membership function and the parameters of the dynamics of the neuro-muscular system which is assumed by 3rd order system.

Some examples of the simulation are shown in Fig. 6. The results of the simulation verify the possibility of constructing a common fuzzy model of the human operator by a set of proposed fuzzy control rules.

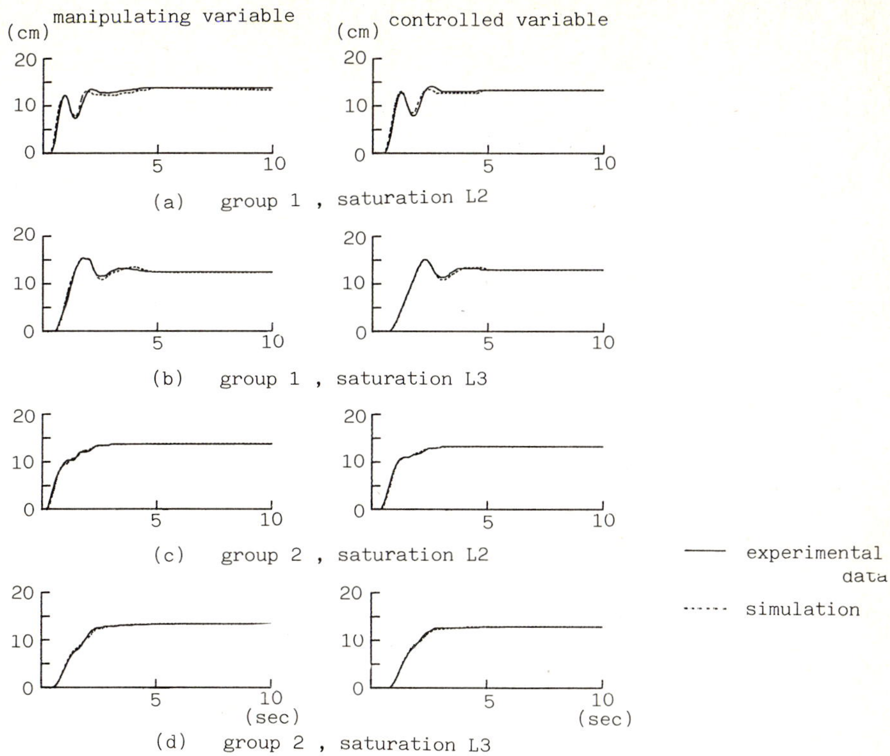

Fig.6 Responses of subjects and model

III-2 Examination of the proposed linguistic control rules with a learning process.

Another experiment was performed in order to examine how the operator's control strategy is changed by the grade of learning in the control task under consideration.

The experiment was carried out by a new subject under the condition of saturation level L3, and the number of trials was 40.

The integral squared error and the settling time are calculated from the experimental data. The result showed that learning of this task finished by about 20 trials.

After the above confirmation, the simulation was executed by using the control rules proposed in III-1.

The result of simulation is shown in Fig. 7.

Fig. 8 shows the grade of membership of cc that was obtained by the simulation.

The result means that the fundamental control strategy during learning process from the initial stage to the steady stage is not changed. The effect of learning is that the grade of the change rate of the controlled variable, cc, which the subject perceives as large is decreased, and the upper bound of cc which is controllable by him becomes higher.

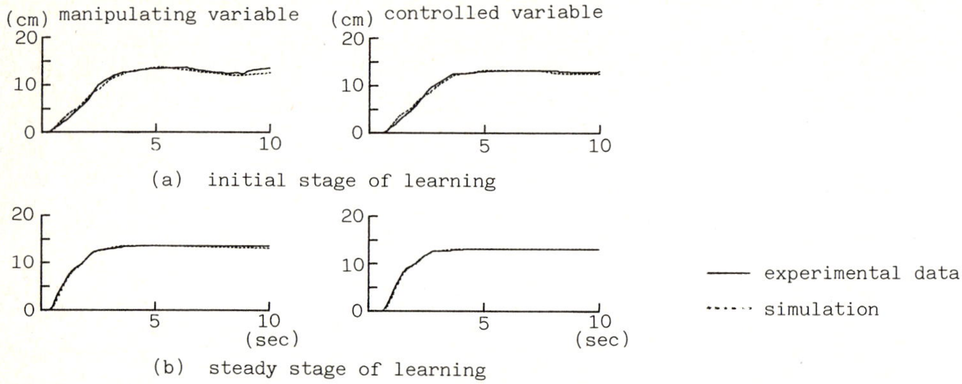

Fig.7 Responses of subject and model in learning stage

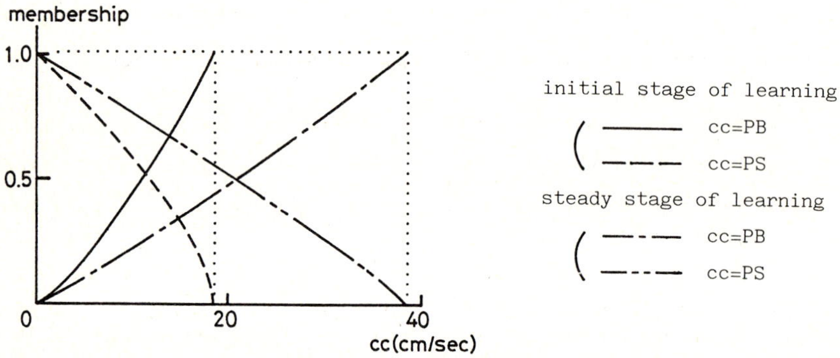

Fig.8 Transition of membership function of the velocity of controlled variable in learning stage

III -3 Reconstruction of the fuzzy model

On the basis of the above-mentioned results, a more natural fuzzy model of the operator may be reconstructed.

In the e-cc plane as shown in Fig. 3, the region where the control rules (1) and (2) contribute to the inference for generating u_c occupies the first quadrant ($e \geq 0$, $cc \geq 0$). This region is further divided into two regions where one control rule dominates more than the other.

The boundary of the two regions is obtained from eqs.(5) and (6) under the conditions $f_1 = f_2$ so $u_c = 0$. The bold line in Fig. 3 shows that boundary.

Furthermore, which state variable dominates the inference is directly obtained from eq.(6) and each region is divided again into two subregions where one state variable dominates more than others. The boundary of two subregions is shown in Fig. 3 as a dashed line.

Actually, the experimental data for increasing u_c exist in region I in Fig. 3, and in region II exist the experimental data for decreasing u_c.

From the above consideration, a new fuzzy model is reconstructed. By this model, the operator's action at each sampling instant is decided by one corresponding control rule.

The set of control rules to be used as the new model is as follows,

(a) if $|e|$ = B then sp = B

or

(b) if $|cc|$ = B then sp = S

for cc $\geq$ 0

(1) if e = PB and if cc = PS then u_c = PB

or

(2) if e = PM and if cc = PM then u_c = PZ or ZR or NZ

or

(3) if cc = PB then u_c = NB

and for cc < 0

(4) if cc = NB then u_c = PVB

or

(5) if cc = NS then u_c = PS

where PM, PZ, ZR, NZ and PVB mean positive medium, positive zero, zero, negative zero and positive very big, respectively.

Some examples of the new model are shown in Fig. 9.

In Fig. 10, as an example, the control rule which was used to generate u_c at each sampling instant is plotted and a trajectory of the controlled variable is drawn in the e-cc plane.

The authors understand that, by this new fuzzy model, each inference of sp and u_c is carried out by each corresponding control rule and therefore this model is a good simulation of the operator's control behavior for the simple task studied here.

IV Conclusion

In this study, a human operator's control behavior in a manual control system with saturation was made clear by constructing a fuzzy model.

Firstly, it was confirmed that the operator's control strategy may be described by a set of linguistic control rules.

Secondly, the grade which each control rule contributes to generating the manipulating variable was made clear.

Lastly, a more natural model of the human operator was proposed and its propriety was verified experimentally.

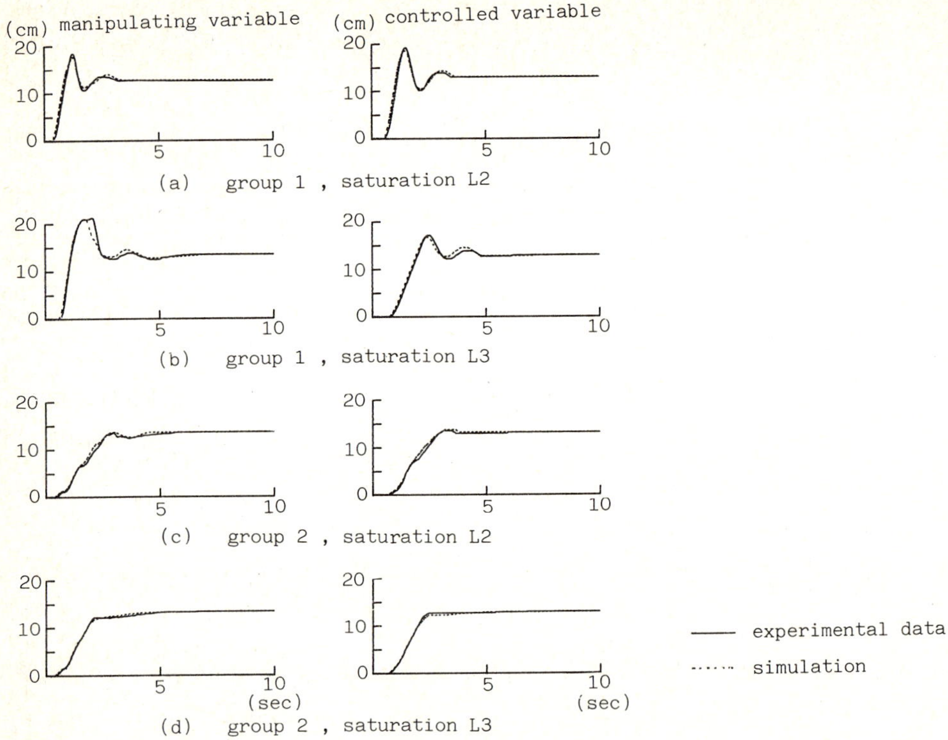

Fig.9 Responses of subjects and reconstructed model

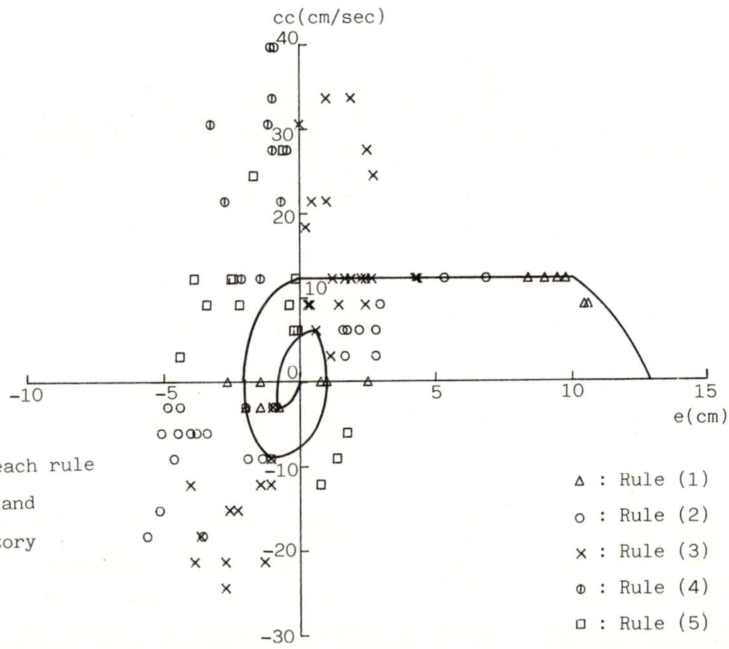

Fig.10

Relation between each rule and generated u_c, and example of trajectory

Acknowledgement

The authors would like to thank for Prof. Dr. H. Terano of Hosei University and Associate prof. Dr. M. Sugeno of Tokyo Institute of Technology, from whom the authors were given many valuable suggestions and encouragement.

Reference

[1] D.McRuer, Human Dynamics in Man-Machine System, Automatica, 16 (1980) 237-253.

[2] T.B.Sheridan and W.R.Ferrell, Man-Machine Systems (MIT Press, 1974).

[3] Y.Maeda, H.Sugiyama and K.Matsushima, On a Fundamental Study of Manual Control for a Controlled Object with Dead Band, Preprint of 21th SICE annual meeting (1982) (in Japanese).

[4] L.A.Zadeh, Theory of Fuzzy Sets, Memorandum No. UCB/ERL M 77-1, University of California, Berkeley (1977).

[5] A.Nagai and K.Matsushima, Remote Mini-manipulator -- Control of its arm and its finger -- , Trans. of SICE, 16-1 (1980) 91-97 (in Japanese).

[6] M.Sugeno, Fuzzy Reasoning, Operations research as a management of science, 26-12 (1981) 692-697 (in Japanese).

Industrial Applications of Fuzzy Control
M. Sugeno (ed.)

THE INFLUENCE OF SOME PARAMETERS ON THE ACCURACY OF A FUZZY MODEL

Jerzy B. Kiszka*, Maria E. Kochańska**, Danuta S. Sliwińska**

* Cybernetics Research Laboratory
College of Engineering
University of Saskatchewan
Saskatoon, Saskatchewan
Canada S7N OWO

** Technical University of Kielce
Al. Tysiaclecia Państwa Polskiego 7,
25-314 Kielce, Poland

A fuzzy model of a d.c. controller and a fuzzy model of a d.c. series motor were built. Root-mean-square error was adopted as the criterion of the model's fit to the real system. The influence of the following parameters on the accuracy of the fuzzy model was investigated: definition of fuzzy implications, interpretation of the sentence connective "also", /and, or/, modification of verbal description, choice of the variables of the system's state, technique for calculation of the determined value of a fuzzy set, and discretization technique for fuzzy set supports. Such a system designer's procedure has been suggested so that the highest possible fit of the fuzzy model to the real object could be achieved.

Key words : fuzzy sets, fuzzy modelling, fuzzy controllers, relational models, model accuracy, electric drive

1. INTRODUCTION

One of the main requirements in mathematical model designs of technological systems is to obtain the highest possible fit of such models in relation to real systems. It is well known that in the classical approach to the construction of mathematical models, an ideal fit is never obtained. As a rule a reasonable compromise is made between the accuracy of a mathematical model and its complexity.

In many situations in which it is impossible to build mathematical models by means of classical techniques, but where the process operator is able to formulate a verbal description of the process behaviour, fuzzy set theory is employed to formalize the verbal description. As a result of such an approach a fuzzy model of the process is obtained. Again the classical question arises about the accuracy and complexity of such a fuzzy model. Of particular importance is the problem of the applicability of a fuzzy model to engineering practice where high precision and quality are required.

There has been a rapid growth of research on fuzzy modelling and fuzzy control since Zadeh first formulated the mathematical foundations of fuzzy model design of complex systems /1973/. Mamdani /1974/ was the first to use a fuzzy algorithm to control the heat engine. A discussion of the current state of research on fuzzy controllers can be found in Mamdani and Sembi /1980/. Willaeys, Malvache and Hammad /1977a/ have presented a method for the construction of a fuzzy controller on the basis of fuzzy model of control process. The concept of a fuzzy algorithm has found application in the control of the velocity of a d.c. motor /Willaeys, Mangin, and Malvache /1977b/, Willaeys, Malvache /1978/, Willaeys, Malvache /1979/: Braae and Rutherford /1979a// have investigated the influence of parameters on the form of a fuzzy controller's model. The linguistic analysis of a fuzzy controller's model is described in Braae and Rutherford /1979b/. The application of self-organizing mechanism to fuzzy control of a multi-dimensional, non-linear process with dead time is presented in Procyk and Mamdani /1979/. Algorithm identification of a fuzzy system has been proposed in Czogała and Pedrycz /1981/. Tong /1979, 1980/ has presented methods for construction of fuzzy models, i.e. verbalisation, fuzzification and identification. He defines four indicators of evaluation quality of a fuzzy model: model complexity, squared error, the absolute difference between discretised, defuzzified model output and the discretised measurement, and a measure of sharpness. He constructed and evaluated fuzzy models for gas furnace data and a river system. In conclusion, Tong stated that an investigation of constructional parameters affecting the model's accuracy should be made in order to obtain a more precise fuzzy model of technological systems.

The purpose of the present paper is to investigate the influence of constructional parameters on the accuracy of a fuzzy model. Section 2 provides a description of a real d.c. power transmission system used in the investigation. Section 3 gives a short outline of the theory of fuzzy model construction. Section 4 contains the results of the experimental investigation. Evaluation of the quality of fuzzy models is carried out by means of a root-mean-square error and by means of a subjective visual assessment of the response curve shape of a fuzzy model. A design algorithm that allows the fuzzy model to attain a high level of accuracy is also given concluding remarks are presented in Section 5.

2. DESCRIPTION OF THE PHYSICAL SYSTEM UNDER INVESTIGATION

A control system for a d.c. series motor /see Fig. 2.1/ is investigated. The task of the control system is to maintain a given constant rotating speed of the motor ω_z despite the effect of the disturbing moment M(t). The series motor is ruled as follows:

Power rating $P_N = 0.7$ KW
Voltage rating $U_N = 110$ V
Rated current $I_N = 8.84$ A
Rated speed $n_N = 1500$ r.p.m.

Set speed is shown in Fig. 2.2. The motor was subjected to disturbances in the form of jump moment of load being equal to a half of rated moment /see Fig. 2.7/. During control procedure the following real quantities were measured: actual velocity ω_r /see Fig. 2.4/, control voltage /see Fig. 2.5/, and motor current /see Fig. 2.6/. From measured error /see Fig. 2.8/, error-sum was determined /see Fig. 2.3/. The control element, which consists of an error detector, integral controller, impulse generator and thyristor chopper /see Fig. 2.1/, was subjected to fuzzy modelling. A fuzzy model of an object consisting of a d.c. series motor and a rate generator was also built. It should be noted that the system under investigation is non-linear, it contains delays and dead time. Furthermore, the system is subjected to stochastic disturbances from the supply system. The responses of the measured real quantities were compared with those of the simulated quantities obtained from the fuzzy models.

3. THE FUZZY MODEL

The process operator, having an assigned control goal, performs visual observation of the process state, control and process output. Next, he intuitively assesses the variables and parameters of the process. From a subjective assessment of those quantities he makes a decision and performs a manual alteration of the system control value so as to be able to accomplish the assigned control goal. In this way he makes a description of the control strategy, a description of process behaviour and finally a subjective operational model of the operator's procedure. It should be noted that such a control algorithm is very flexible and adequate to the actual process situation, and as a rule it is much better than a control algorithm obtained by using modern control theory. The reason for such a situation seems to be very simple: the linguistic algorithm of control comprises all "metaphysical" skills of the operator, such as intuition, experience, intelligence and so on, which cannot be dealt with by modern mathematics. The operator's "mind" and his knowledge of the controlled technological process can be formalized mathematically using the theory of fuzzy sets conceived by L. A. Zadeh /1973/.

Suppose that in observing an error and error sum, the process operator has made a manual control of the process according to this own hypothetical verbal description:

IF E = big AND S = small THEN U = zero
ALSO
IF E = medium AND S = zero THEN U = big
ALSO
. (3.1)
.
.
ALSO
IF E = zero AND S = zero THEN U = zero

Let

$$\mathcal{E} = \{e_i\} \quad (-\infty, \infty)$$
$$\mathcal{S} = \{s_j\} \quad (-\infty, \infty) \tag{3.2}$$
$$\mathcal{U} = \{u_k\} \quad (-\infty, \infty)$$

be the finite space of error, error sum and control.

Let

$$\mathcal{F}(E) = \{E \mid \mathcal{E} \rightarrow [0, 1]\}$$
$$\mathcal{F}(S) = \{S \mid \mathcal{S} \rightarrow [0, 1]\} \tag{3.3}$$
$$\mathcal{F}(U) = \{U \mid \mathcal{U} \rightarrow [0, 1]\}$$

be the finite family of fuzzy set of error, error-sum and control, then the above-presented verbal description (3.1) can be formalized in the formula:

$$R : \mathcal{F}(E) \times \mathcal{F}(S) \rightarrow \mathcal{F}(U) \tag{3.4}$$

where:

x - fuzzy Cartesian product,

$\rightarrow$ - fuzzy implication

which is a so-called fuzzy relation /L. A. Zadeh, 1973/. In order to calculate the control U, when given the fuzzy relation R, error E and error-sum S, the compositional rule of inference is to be used /L. A. Zadeh, 1973/ :

$$U = E \circ S \circ R \tag{3.5}$$

where: o - max-min composition.

Since in the on-line control system a non-fuzzy control action is necessary, one must perform defuzzification of a fuzzy set U calculated from formula (3.5). An extensive discussion of defuzzification operators can be found in Braae and Rutherford /1978/. In practical applications, most frequently error and error-sum are available in the determined form. Therefore, it is more convenient to use fuzzy controller in the form of the so-called look-up table /see Rutherford and Bloore /1976//.

It should be noted that there are great possibilities for modifying the performance of a fuzzy model, particularly its fit by means of alterations to the verbal description (3.1), proper choice of discretization of the space of the process state variables (3.2), proper choice of definition of the fuzzy implication of expression (3.4), proper interpretation of the sentence connective "also" in the description (3.1), another definition of composition of fuzzy sets in a formula (3.5), effective choice of the operator of the defuzzification of an output fuzzy set, proper choice of the process state variables. The problem of modifying the quality of a fuzzy model by means of the above-listed factors will be the object of our investigation. We shall adopt the root-mean-square error and subjective visual assessment of the response as criteria for assessing fuzzy model quality.

4. EXPERIMENTAL STUDIES

The process operator observes the courses of the real variables of the process described in Section 2 and tries to formulate verbal descriptions of the process behaviour as faithfully as possible. Verbal descriptions of the process are processed by a CDRA 1305 computer according to the theoretical outline presented in Section 3. The real variables of the control process obtained through measurements are fed into the computer. The experiments were carried in two stages /see Fig. 4.1/. In the first stage, fuzzy relations were determined from the verbal description and subsequently the look-up table was made in which are encoded answers of the system to every possible combination of input signals. In the second stage, testing was made on the basis of real input signals; theoretical answers are read from the look-up table and are compared with real answers.

In order to save the computer's memory the investigators did not construct a complete relational table. Only fuzzy sets being answers to subsequent possible input singletons were constructed. For each set of that type the determined value is given and stored in memory. For example, in the case of verbal descriptions in the form IF E AND S THEN U, the operating vector requires for the determination of the fuzzy relation table contains

ME * MS + MU

numerical values, while explicit determination of the relational table would have required

(1 + LOP) * (ME * MS * MU) + ME * MS

numerical values, where:

ME = card $\mathcal{E}$

MS = card $\mathcal{S}$

MU = card $\mathcal{U}$

are cardinals of fuzzy set supports for E, S, and U, respectively, and LOP - the number of sentences in the verbal description.

It should be noted that such a technique of formulating the fuzzy relation table is very useful for practical applications.

The quality of a fuzzy model was estimated on the basis of the root-mean-square error

$$\Delta^2 e = \frac{\sum_{i=1}^{n} (ur_i - um_i)^2}{\sum_{i=1}^{n} ur_i^2} \qquad (4.1)$$

where:

ur_i - answer of the real system at time i,

um_i - answer of the fuzzy model at time i,

n - number discretization intervals of time axis.

Also a visual assessment of the fuzzy model quality was made by observing computer printouts of the curves obtained from the fuzzy model and the real system. The time to each characteristic points by the answer of the real system and that of the model was calculated. Convergence rate of the model's answer to the real curve was estimated. Instantaneous deviation quantity between the real system's answer and the model's answer in some selected time moments

$$\Delta e_i = ur_i - um_i$$

was calculated. Finally, a general visual assessment of the shape of the model's answer curve was made and it was classified as very bad, bad, satisfactory, good, very good.

Experiment I

The process operator observes the course of error-sum S and the course of control U /see Figs. 2.3 and 2.5/ and on the basis of his observations he creates a verbal description of the control element in the form:

	1/	IF S = null	THEN U = null
also			
	2/	IF S = null	THEN U = zero
also			
	3/	IF S = null	THEN U = big
also			
	4/	IF S = null OR zero	THEN U = zero
also			
	5/	IF S = null OR zero	THEN U = medium
also			
	6/	IF S = null OR zero	THEN U = medium OR big
also			
	7/	IF S = zero	THEN U = null
also			
	8/	IF S = zero	THEN U = zero
also			
	9/	IF S = zero	THEN U = zero OR small
also			
	10/	IF S = zero OR small	THEN U = zero
also			
	11/	IF S = small	THEN U = zero
also			
	12/	IF S = small OR medium	THEN U = null
also			
	13/	IF S = small OR medium	THEN U = null OR zero
also			
	14/	IF S = small OR medium	THEN U = small
also			
	15/	IF S = medium	THEN U = medium
also			
	16/	IF S = big	THEN U = small
also			
	17/	IF S = big OR super-big	THEN U = zero OR small
also			

18/ IF S = super-big THEN U = zero

also

19/ IF S = negative null THEN U = zero

also

20/ IF S = negative null OR negative zero THEN U = big

also

21/ IF S = negative small THEN U = super-big

also

22/ IF S = negative small OR negative medium THEN U = null OR zero

also

23/ IF S = negative big OR negative medium THEN U = zero

also

24/ IF S = negative big OR negative super-big THEN U = small

also

25/ IF S = negative big OR negative super-big THEN U = big

also

26/ IF S = negative super-big THEN U = medium OR big

The definitions of fuzzy sets for variables S and U are shown in Tables 4.1 and 4.2.

It should be noted that fuzzy sets subjectively defined are symmetrical and are intersected at the point where the membership functions reach the value $\mu = 0.5$. Six fuzzy sets were suggested for control U and twelve fuzzy sets for error-sum S. Discretisation of fuzzy set support was made for control U at thirty-one points, but where discretisation of fuzzy set support was made for error-sum S at twenty-five points /see Tables 4.1 and 4.2/. The definitions of fuzzy implication: IF S = A THEN U = B was adopted according to the formula:

$$\mu_{SU}(s, u) = \min(\mu_A(s), \mu_B(u)) \qquad (4.2)$$

The sentence connective "also" was interpreted as "or" - the union of fuzzy sets. The classical definition of "max-min" composition of fuzzy relations in the compositional rule of inference was adopted. The experiment was carried out in two versions differing from each other by the technique of calculating the deterministic value of fuzzy set U:

Ia/ according to the criterion max

$$u_o \in \{\bar{u}: \mu_U(\bar{u}) = \max_u \mu_U(u)\} \qquad (4.3)$$

In case when membership function $\mu_U(u)$ reached the maximum value for several different values of u, the first encountered value of u for which $\mu_U(u)$ reached the max value, was taken as the deterministic value.

Ib/ according to the criterion of root-mean-square value

$$u_o = \frac{\sum_{i=1}^{n} u_i}{n} \qquad (4.4)$$

where:

u_i - the support values in which function $\mu_U(u)$ reaches max

n - number of support elements in which membership function $\mu_U(u)$ reaches the max value

The experimental results in which the deterministic value was calculated according to formula (4.3) are presented in Fig.4.2. The experimental results in which the deterministic value was calculated according to formula (4.4) are presented in Fig. 4.3. Looking at the results of the experiments, we notice a lack of fit between the real system and the fuzzy model - a big root-mean-square error. It is also noteworthy that in both the above cases fuzzy control processes are unconvergent - they do not reach the assigned value.

In the analysis of the experimental results, the following hypotheses for lack of fit of the fuzzy model were made:

1/ Wrong choice of the control process state variables as the cause of the fact that the verbal description given above do not fully characterize the investigated phenomenon. Fig. 4.4 represents the deterministic dynamic phase characteristics of the control process of a series motor S = f(U) and I = f(U). On the basis of this diagram one can conclude that the verbal description contains ambiguities because one value of error-sum S corresponds, in most cases, to more than one control value U.

2/ The way of calculating the deterministic value has an essential significance on the adequacy of a fuzzy model and thus in experiment Ia, root-mean-square error $\Delta^2 e = 0.259$, but where in experiment Ib, root-mean-square error $\Delta^2 e = 0.077$.

On the basis of the above discussion the following experiment was performed, introducing two corrections:

1/ to take into account an additional variable of the process state - motor current I.

2/ to calculate deterministic value of a fuzzy set U according to formula (4.4).

Eeperiment II

In observing the control process and taking into account corrections from Experiment I, the operator formulated the following verbal description of the control element:

1/ IF S = null AND I = small THEN U = zero

also

2/ IF S = null AND I = big THEN U = big

also

3/ IF S = zero AND I = small THEN U = zero

also

4/ IF S = zero AND I = big THEN U = big OR medium

also

5/ IF S = small AND I = small THEN U = null

also

6/ IF S = small AND I = medium THEN U = zero

also

7/ IF S = small AND I = big THEN U = null

also

8/ IF S = medium AND I = big THEN U = small

also

9/ IF S = big AND I = medium THEN U = null

also

10/ IF S = super-big AND I = medium THEN U = zero

also

11/ IF S = negative null AND I = zero THEN U = null

also

12/ IF S = negative zero AND I = medium THEN U = big

also

13/ IF S = negative small AND I = small THEN U = big

also

14/ IF S = negative medium AND I = small THEN U = big

also

15/ IF S = negative big AND I = small THEN U = zero

also

16/ IF S = negative super-big AND I = small THEN U = zero

Definitions of fuzzy sets for variables S, I, and U are shown in Tables 4.3, 4.4 and 4.5. It should be noted that the way of discretisation of fuzzy set supports was changed, for variables S and variable U. Discretisation of fuzzy set support of error-sum in 23 points and discretisation of fuzzy set support of control U in 24 points was made /compare Tables 4.1 and 4.2/.

The experiment was performed for the three following definitions of a fuzzy relation:

IIa/ $$\mu_R(s, i, u) = \vee \{1 \wedge [1 - (\mu_S(s) \wedge \mu_I(i)) + \mu_U(u)]\} \qquad (4.5)$$

IIb/ $$\mu_R(s, i, u) = \wedge \{1 \wedge [1 - (\mu_S(s) \wedge \mu_I(i)) + \mu_U(u)]\} \qquad (4.6)$$

IIc/ $$\mu_R(s, i, u) = \vee \{\mu_S(s) \wedge \mu_I(i) \wedge \mu_U(u)\} \qquad (4.7)$$

where: $\vee$ - max, $\wedge$ - min

The deterministic values of fuzzy set U were calculated according to formula (4.4). The experimental results are shown in Figs. 4.5, 4.6, 4.7 and Tables 4.6 and 4.7.

From the results obtained, one can formulate the following conclusions:

1/ Introduction of the definition of fuzzy implications according to Lukasiewicz's logic IIa and the interpretation of the sentence connective "also" as union /see formula (4.5)/ did not give expected results /see Fig. 4.5 and Table 4.6/.

2/ Interpretation of the sentence connective "also" as intersection for the same definition of implication /see formula (4.6)/ considerably improved the result obtained /see Fig. 4.6 and Table 4.7/.

3/ Application of the definition of implication according to L. A. Zadeh /see formula (4.7)/ gave results similar to those obtained in Experiment IIb /see Fig.4.7/. However, the number of computations was significantly smaller.

4/ It is noteworthy that the control processes in Experiment IIa are unconvergent, but where in Experiments IIb and IIc, they are very poorly convergent.

5/ A decreasing tendency of root-mean-square error is seen. However general visual assessment of the response runs of the simulated process is insufficient. High instantaneous values of deviations between the response of control element and the response of modelled fuzzy controller are observed.

6/ It should be also noted that in Experiment I, the number of control rules was 26 and the verbal description was a form of simple implication IF...THEN... . On the other hand, in Experiment II, there were 16 control rules and each verbal description has the form of composite implication IF...AND...THEN... .

On the basis of the above discussion another experiment was performed, introducing the following corrections:

1/ extended verbal descriptions of the process so as to eliminate the above-described negative phenomena, mainly by increasing the number of sentences /number of control rules/.

2/ to adopt for further calculations the definition of a fuzzy relation given by L. A. Zadeh /see formula 4.7/ because of the smaller number of mathematical operations that must be performed.

Experiment III

Taking into account the above remarks, the process operator formulated the following verbal description of control process /control element/:

1/ IF S = null AND I = small THEN U = zero

also

2/ IF S = null AND I = big THEN U = big

also

3/ IF S = zero AND I = small THEN U = zero

also

4/ IF S = zero AND I = big THEN U = medium

also

5/ IF S = small AND I = zero THEN U = null

also

6/ IF S = small AND I = small THEN U = null

also

7/ IF S = small AND I = medium THEN U = zero

also

8/ IF S = small AND I = big THEN U = zero

also

9/ IF S = medium AND I = small THEN U = null

also

10/ IF S = small AND I = big THEN U = small

also

11/ IF S = big AND I = medium THEN U = zero

also

12/ IF S = big AND I = big THEN U = small

also

13/ IF S = super-big AND I = medium THEN U = zero

also

14/ IF S = negative null AND I = zero THEN U = null

also

15/ IF S = negative null AND I = medium THEN U = zero

also

16/ IF S = negative zero AND I = zero THEN U = null

also

17/ IF S = negative zero AND I = small THEN U = zero

also

18/ IF S = negative zero AND I = medium THEN U = big

also

19/ IF S = negative small AND I = zero THEN U = zero

also

20/ IF S = negative small AND I = small THEN U = big

also

21/ IF S = negative medium AND I = zero THEN U = null

also

22/ IF S = negative medium AND I = small THEN U = big

also

23/ IF S = negative medium AND I = medium THEN U = big

also

24/ IF S = negative big AND I = zero THEN U = zero

also

25/ IF S = negative big AND I = small THEN U = big

also

26/ IF S = negative super-big AND I = zero THEN U = small

also

27/ IF S = negative super-big AND I = small THEN U = medium

Fuzzy sets of variables S, I, and U were defined in the same way as in Experiment II. The fuzzy relation was determined according to formula (4.7). Deterministic values of fuzzy sets U were calculated according to formula (4.4). The results of the experiment are represented in Fig. 4.8 and Table 4.8.

From the above results, one can formulate the following conclusions:

1/ A fuzzy model reproduces the real control process with sufficient accuracy - root-mean-square error $\Delta^2 e = 0.00747$

2/ The convergence of fuzzy control process is considered to be good.

3/ The characteristic points of both curves - theoretical and real ones, i.e. extremes and final values, are superimposed.

4/ A high value of instantaneous deviation Δe_i is seen between the real control value and the control value obtained from fuzzy logic controller on the segment $|\alpha - \alpha|$ of control curves /see Fig. 4.8/. Note that it is just on this segment $|\alpha - \alpha|$ that both error-sum S and motor current I are slowly changing, but where control U is quickly changing.

5/ The number of control rules /sentences/ in the verbal description has increased up to 27 and is much greater than in Experiment II.

6/ One can assume that the holistic visual assessment of the theoretical curve shape is good. From the above remarks, one can state that a fuzzy model obtained in this way for the control element is good and from the point of view of requirements made for simulation methods of technological devices is applicable in practice. Particular attention should be given to the fact that in most experiments a very small root-mean-square error $\Delta^2 e$ was obtained. It is obvious that the creator of this phenomenon is the process operator who tries to reproduce the real control process as faithfully as possible. His attempts to obtain a precise model are encoded in the verbal description of the control process and the root-mean-square error is a measure of this endeavour.

Experiment IV

Since the results of Experiment III were satisfactory, a fuzzy model of a d.c. series motor was built /see Fig. 2.1/. Control U was adopted as the input variable, current intensity I - as the state variable and motor rotations Ω - as the output variable.

The process operator gave the following verbal description of motor performance:

1/ IF U = null AND I = zero THEN Ω = medium

also

2/ IF U = null AND I = small THEN Ω = small

also

3/ IF U = null AND I = medium THEN Ω = zero

also

4/ IF U = zero AND I = small THEN Ω = medium

also

5/ IF U = zero AND I = zero THEN Ω = big

also

6/ IF U = zero AND I = medium THEN Ω = zero

also

7/ IF U = zero AND I = big THEN Ω = medium

also

8/ IF U = small AND I = small THEN Ω = super-big

also

9/ IF U = small AND I = big THEN Ω = small

also

10/ IF U = medium AND I = medium THEN Ω = super-big

also

11/ IF U = medium AND I = big THEN Ω = medium

also

12/ IF U = big AND I = small THEN Ω = super-big

also

13/ IF U = big AND I = medium THEN Ω = big

also

14/ IF U = big AND I = big THEN Ω = medium

also

15/ IF U = super-big AND I = medium THEN Ω = big

also

16/ IF U = super-big AND I = small THEN Ω = big

The definition of the fuzzy sets for variables U, I and Ω are presented in Tables 4.9, 4.10, 4.11. The same mathematical apparatus was used as in Experiment III. The experimental results are shown in Fig. 4.9.

Estimating the experimental results, one must state that the fuzzy model of the d.c. motor is good. The root-mean-square error is very small and equals $\Delta^2 e = 0.006292$.

The instantaneous value of the error deviation of the model Δe_i in particular time moments is small. The instantaneous value of the error deviation Δe_i changes sign eight times. The time to reach the set speed value ω_z = 2000 r.p.m. is satisfactory and equals t_u = 54 seconds. The fuzzy model shows a very good property of convergence. It should be noted that the number of sentences in the verbal description is relatively small and equals 16 and nevertheless the adequacy and quality of the motor's fuzzy model is good.

5. A SYSTEM DESIGNER'S PROCEDURE

The above experiments show that using the algorithm of the system designer's procedure one can obtain a fuzzy model of the investigated phenomenon with the possibly of high quality, applicable to the investigation of properties and to the design of technological installations.

Suppose that the system designer received the original verbal description constructed by the operator /see Fig. 5.1/. In order to obtain the highest possible quality of the fuzzy model, the designer should:

1/ make an adequate choice of the process state variables; done by way of an interview with the operator and observations of the process; he should be guided by the principle that the number of state variables is to be equal to the number of independent energy stores of the process;

2/ chose the variability ranges /supports/ of the state variables and make their adequate discretisation;

3/ define the fuzzy sets for fuzzy variables;

4/ chose an appropriate mathematical apparatus:
   a/ formulate definitions of fuzzy implications,
   b/ make an appropriate interpretation of the sentence connective "also",
   c/ chose an appropriate technique for calculating deterministic values of the output fuzzy sets,
   d/ chose the technique for calculating the look-up table.

5/ perform an analysis of the fuzzy model through:
   a/ calculation of the root-mean-square error $\Delta^2 e$,
   b/ estimation of momentary value of the model error deviation Δe_i,

c/ estimation of the degree of model convergence,

d/ estimation of the time to reach characteristic points by the response of the fuzzy model.

In the case where the estimation of the model's quality is negative, the system designer should:

6/ perform a modification of the verbal description through:

a/ an increase or decrease in the number of sentences in the verbal description,

b/ a change to the values of the fuzzy variables in the fuzzy implication.

Next he should repeat the whole procedure beginning with point 1/.

In the case where the estimation of the model's quality is satisfactory, the system designer should:

7/ apply the fuzzy model obtained to the testing of the properties of a given phenomenon or to the construction of technological installations.

The investigations performed confirmed the accuracy of the procedure proposed in Fig. 5.1.

6. CONCLUDING REMARKS

It follows from the investigation performed that the constructional parameters of a fuzzy model give a high possibility of exerting influence on its quality. An appropriate choice of the model's parameters by the system designer causes the fuzzy model to achieve high accuracy.

It seems that high possibilities of further shaping the quality of the fuzzy model exist in an appropriate for fuzzy implications; in an appropriate choice of operator for composing fuzzy relations /compositional rule of inference/; and in an appropriate choice of operator for defuzzification of output fuzzy sets.

The general conclusion from the investigations performed is that the fuzzy model is fully applicable to the investigation of power installations -electrical drives, where high accuracy and quality of modelling are required.

Apart from high accuracy and quality of the fuzzy model, the following advantages should be emphasized: simplicity of construction, the designer need have little knowledge of the investigated phenomenon, small use of the computer's memory, short time of performing mathematical operations, a feasibility of describing phenomena that are difficult to deal with by classical methods.

REFERENCES

[1] M. Braae, D. A. Rutherford: Fuzzy relations in a control setting, Kybernetes, 2, 1978

[2] M. Braae, D. A. Rutherford: Selection of parameters for a fuzzy logic controller, Fuzzy Sets and Systems, 2, 1979a

[3] M. Braae, D. A. Rutherford: Theoretical and linguistic aspects of the fuzzy logic controller, Automatica, 15, 1979b

[4] D. Czogala, W. Pedrycz: On identification in fuzzy systems and its applications in control problems, Fuzzy Sets and Systems, 6, 1981

[5] E. H. Mamdani: Application of fuzzy algorithms for control of simple dynamic plant, Proceedings of the Institution of Electrical Engineers, Control and Science, 121, 12, December 1974

[6] E. H. Mamdani, B. S. Sembi: Process control using fuzzy logic, in: Fuzzy Sets - Theory and Applications to Policy Analysis and Information Systems, P. P. Wang, S. K. Chang (eds.), Plenum Press, New York, 1980

[7] T. J. Procyk, E. H. Mamdani: A linguistic self-organizing process controller, Automatica, 15, 1979

[8] D. A. Rutherford, G. C. Bloore: The implementation of fuzzy algorithms for control, Proceedings of the IEEE, April 1976

[9] R. M. Tong: The construction and evaluation of fuzzy models, in: Advances in Fuzzy Set Theory and Applications, M. M. Gupta, R. K. Rapade, R. R. Yager (eds.), North-Holland Publishing Company, 1979

[10] R. M. Tong: The evaluation of fuzzy models derived from experimental data, Fuzzy Sets and Systems, 4, 1980

[11] D. Willaeys, N. Malvache, P. Hammad: Utilization of fuzzy sets for system modelling and control, IEEE Conference on Decision and Control, New Orleans, December 1977a

[12] D. Willaeys, P. Mangin, N. Malvache: Use of fuzzy sets for systems modelling and control: application to the speed repulation of a strongly perturbed motor, in: Digital Computer Applications to Process Control, H. R. Van Nauta Lemke (ed.), North-Holland Publishing Company, 1977b

[13] D. Willaeys, N. Malvache: Use of fuzzy model for process control, Proceedings of the International Conference on Cybernetics and Society, New York, 1978

[14] D. Willaeys, N. Malvache: Contribution of the fuzzy set theory to man-machine system, in: Advances in Fuzzy Set Theory and Applications, M. M. Gupta, R. K. Rapade, R. R. Yager (eds.), North-Holland Publishing Company, 1979

[15] L. A. Zadeh: Outline of a new approach to the analysis of complex systems and decision processes, IEEE Transaction on Systems, Man and Cybernetics, SMC-3, 1973

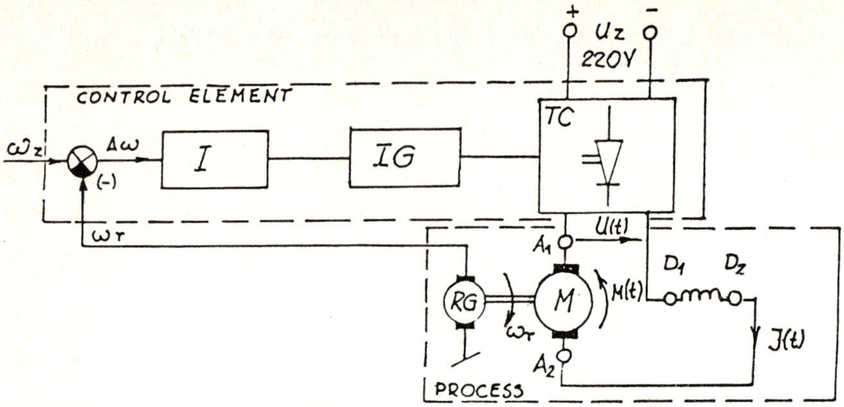

I - integral controller

IG - impulse generator

TC - thyristor chopper

M - series motor

RG - rate generator

D_1 D_2 - excitation winding

ω_z - setpoint

$\Delta\omega$ - control error

ω_r - actual velocity

u_z - supply voltage

u(t) - control voltage

J(t) - motor current

M(t) - disturbing moment

Fig. 2.1 Electric drive

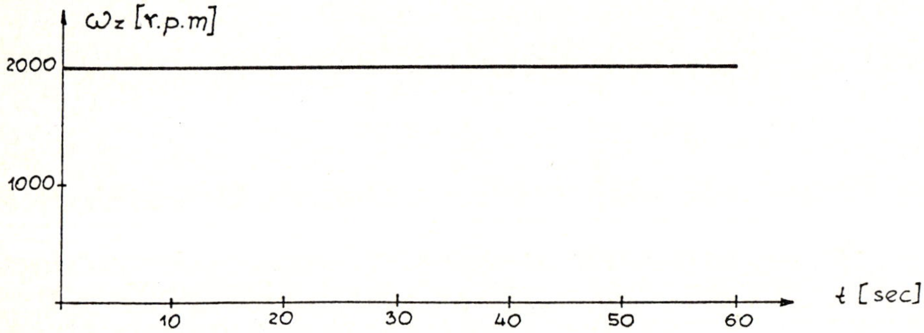

Fig. 2.2 Set point

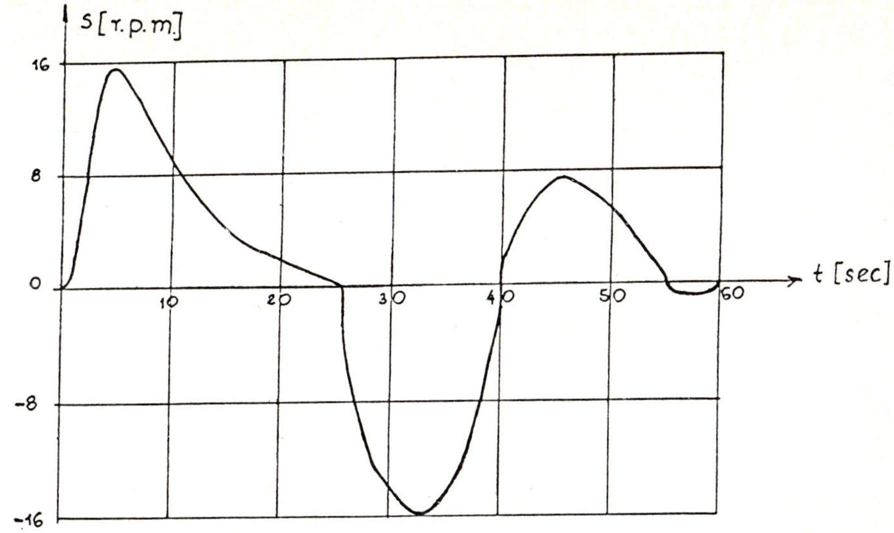

Fig. 2.3 Error - sum

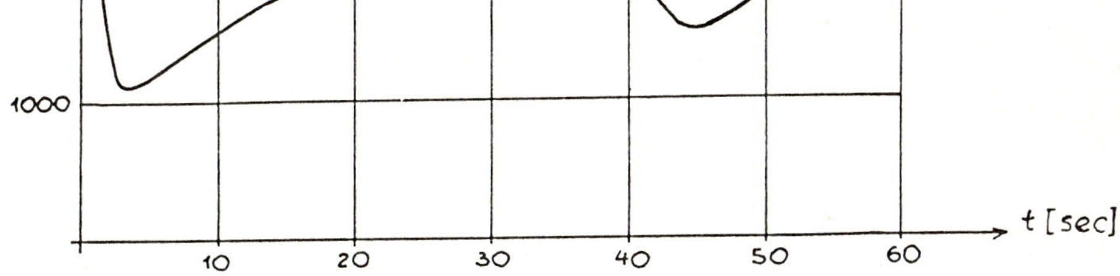

Fig. 2.4 Actual velocity

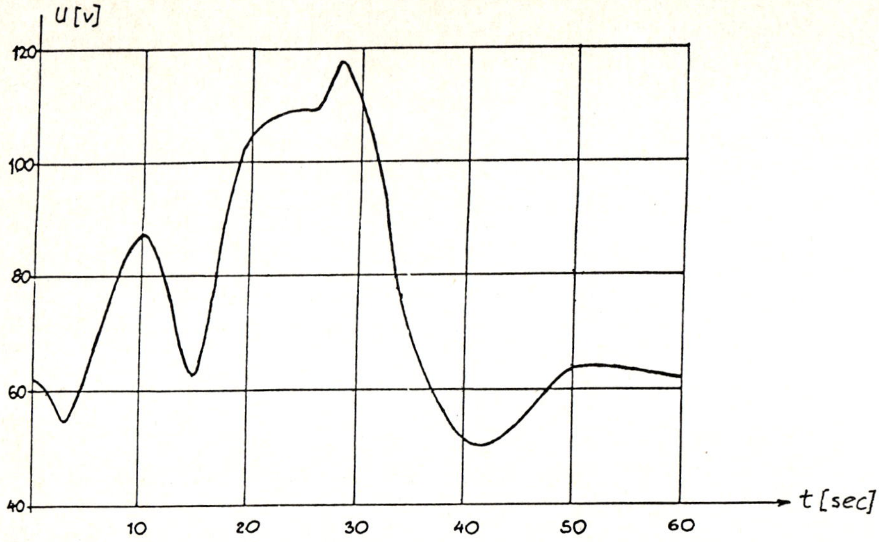

Fig. 2.5 Control voltage

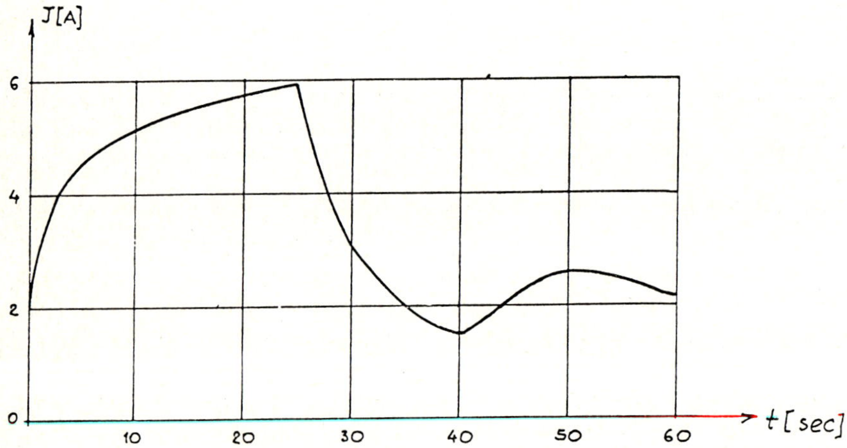

Fig. 2.6 Motor current

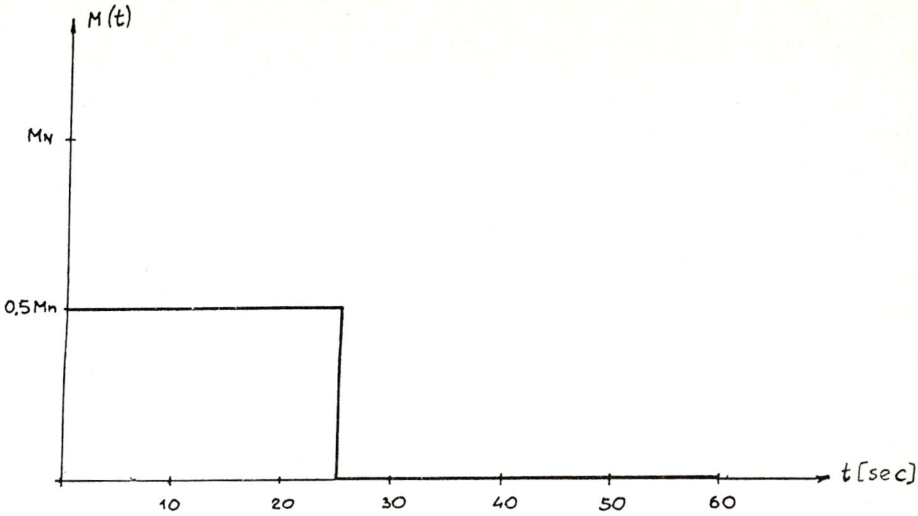

M_N - nominal moment

Fig. 2.7 Disturbing moment

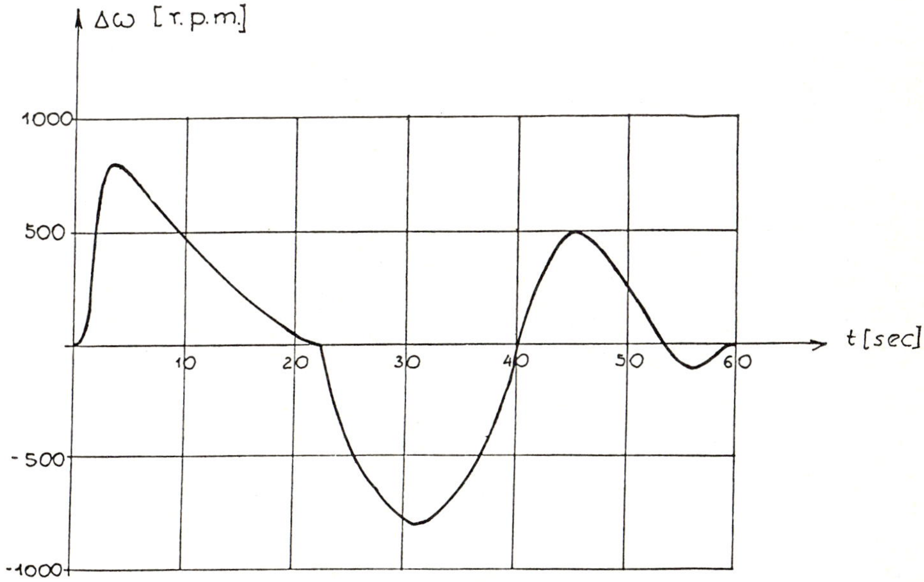

Fig. 2.8 Error

ERROR - SUM S

S universe \ fuzzy sets	NSB	NB	NM	NS	NZE	NN	PN	PZE	PS	PM	PB	PSB
-17.0	1.	0.	0.	0.	0.	0.	0.	0.	0.	0.	0.	0.
-16.0	1.	0.	0.	0.	0.	0.	0.	0.	0.	0.	0.	0.
-15.0	0.75	0.	0.	0.	0.	0.	0.	0.	0.	0.	0.	0.
-13.5	0.5	0.5	0.	0.	0.	0.	0.	0.	0.	0.	0.	0.
-12.0	0.	1.	0.	0.	0.	0.	0.	0.	0.	0.	0.	0.
-10.5	0.	0.5	0.5	0.	0.	0.	0.	0.	0.	0.	0.	0.
-9.0	0.	0.	1.	0.	0.	0.	0.	0.	0.	0.	0.	0.
-7.5	0.	0.	0.5	0.5	0.	0.	0.	0.	0.	0.	0.	0.
-6.0	0.	0.	0.	1.	0.	0.	0.	0.	0.	0.	0.	0.
-4.5	0.	0.	0.	0.5	0.5	0.	0.	0.	0.	0.	0.	0.
-3.0	0.	0.	0.	0.	1.	0.	0.	0.	0.	0.	0.	0.
-1.5	0.	0.	0.	0.	0.5	0.5	0.	0.	0.	0.	0.	0.
0.0	0.	0.	0.	0.	0.	1.	1.	0.	0.	0.	0.	0.
1.5	0.	0.	0.	0.	0.	0.	0.5	0.5	0.	0.	0.	0.
3.0	0.	0.	0.	0.	0.	0.	0.	1.	0.	0.	0.	0.
4.5	0.	0.	0.	0.	0.	0.	0.	0.5	0.5	0.	0.	0.
6.0	0.	0.	0.	0.	0.	0.	0.	0.	1.	0.	0.	0.
7.5	0.	0.	0.	0.	0.	0.	0.	0.	0.5	0.5	0.	0.
9.0	0.	0.	0.	0.	0.	0.	0.	0.	0.	1.	0.	0.
10.5	0.	0.	0.	0.	0.	0.	0.	0.	0.	0.5	0.5	0.
12.0	0.	0.	0.	0.	0.	0.	0.	0.	0.	0.	1.	0.
13.5	0.	0.	0.	0.	0.	0.	0.	0.	0.	0.	0.5	0.5
15.0	0.	0.	0.	0.	0.	0.	0.	0.	0.	0.	0.	0.75
16.0	0.	0.	0.	0.	0.	0.	0.	0.	0.	0.	0.	1.
17.0	0.	0.	0.	0.	0.	0.	0.	0.	0.	0.	0.	1.

NSB - negative super-big
NB - negative big
NM - negative medium
NS - negative small
NZE - negative zero
NN - negative null

PN - positive null
PZE - positive zero
PS - positive small
PM - positive medium
PB - positive big
PSB - positive super-big

Table 4.1 Error-sum fuzzy sets for Experiment I

CONTROL VARIABLE U

U universe \ fuzzy sets	N	ZE	S	M	B	SB
50.0	1.	0.	0.	0.	0.	0.
52.5	0.83	0.17	0.	0.	0.	0.
55.0	0.67	0.33	0.	0.	0.	0.
57.5	0.5	0.5	0.	0.	0.	0.
60.0	0.33	0.67	0.	0.	0.	0.
62.5	0.17	0.83	0.	0.	0.	0.
65.0	0.	1.	0.	0.	0.	0.
67.5	0.	0.83	0.17	0.	0.	0.
70.0	0.	0.67	0.33	0.	0.	0.
72.5	0.	0.5	0.5	0.	0.	0.
75.0	0.	0.33	0.67	0.	0.	0.
77.5	0.	0.17	0.83	0.	0.	0.
80.0	0.	0.	1.	0.	0.	0.
82.5	0.	0.	0.83	0.17	0.	0.
85.0	0.	0.	0.67	0.33	0.	0.
87.5	0.	0.	0.5	0.5	0.	0.
90.0	0.	0.	0.33	0.67	0.	0.
92.5	0.	0.	0.17	0.83	0.	0.
95.0	0.	0.	0.	1.	0.	0.
97.5	0.	0.	0.	0.83	0.17	0.
100.0	0.	0.	0.	0.67	0.33	0.
102.5	0.	0.	0.	0.5	0.5	0.
105.0	0.	0.	0.	0.33	0.67	0.
107.5	0.	0.	0.	0.17	0.83	0.
110.0	0.	0.	0.	0.	1.	0.
112.5	0.	0.	0.	0.	0.83	0.17
115.0	0.	0.	0.	0.	0.67	0.33
117.5	0.	0.	0.	0.	0.5	0.5
120.0	0.	0.	0.	0.	0.33	0.67
122.5	0.	0.	0.	0.	0.17	0.83
125.0	0.	0.	0.	0.	0.	1.

N - null, ZE - zero, S - small, M - medium, B - big, SB - super big

Table 4.2 Control variable fuzzy sets for Experiment I

ERROR - SUM S

S universe \ fuzzy sets	NSB	NB	NM	NS	NZE	NN	PN	PZE	PS	PM	PB	PSB
-16.0	1.	0.	0.	0.	0.	0.	0.	0.	0.	0.	0.	0.
-15.0	0.75	0.	0.	0.	0.	0.	0.	0.	0.	0.	0.	0.
-13.5	0.35	0.5	0.	0.	0.	0.	0.	0.	0.	0.	0.	0.
-12.0	0.	1.	0.	0.	0.	0.	0.	0.	0.	0.	0.	0.
-10.5	0.	0.5	0.5	0.	0.	0.	0.	0.	0.	0.	0.	0.
-9.0	0.	0.	1.	0.	0.	0.	0.	0.	0.	0.	0.	0.
-7.5	0.	0.	0.5	0.5	0.	0.	0.	0.	0.	0.	0.	0.
-6.0	0.	0.	0.	1.	0.	0.	0.	0.	0.	0.	0.	0.
-4.5	0.	0.	0.	0.5	0.5	0.	0.	0.	0.	0.	0.	0.
-3.0	0.	0.	0.	0.	1.	0.	0.	0.	0.	0.	0.	0.
-1.5	0.	0.	0.	0.	0.5	0.5	0.	0.	0.	0.	0.	0.
0.0	0.	0.	0.	0.	0.	1.	1.	0.	0.	0.	0.	0.
1.5	0.	0.	0.	0.	0.	0.	0.5	0.5	0.	0.	0.	0.
3.0	0.	0.	0.	0.	0.	0.	0.	1.	0.	0.	0.	0.
4.5	0.	0.	0.	0.	0.	0.	0.	0.5	0.5	0.	0.	0.
6.0	0.	0.	0.	0.	0.	0.	0.	0.	1.	0.	0.	0.
7.5	0.	0.	0.	0.	0.	0.	0.	0.	0.5	0.5	0.	0.
9.0	0.	0.	0.	0.	0.	0.	0.	0.	0.	1.	0.	0.
10.5	0.	0.	0.	0.	0.	0.	0.	0.	0.	0.5	0.5	0.
12.0	0.	0.	0.	0.	0.	0.	0.	0.	0.	0.	1.	0.
13.5	0.	0.	0.	0.	0.	0.	0.	0.	0.	0.	0.5	0.35
15.0	0.	0.	0.	0.	0.	0.	0.	0.	0.	0.	0.	0.75
16.0	0.	0.	0.	0.	0.	0.	0.	0.	0.	0.	0.	1.

Table 4.3 Error-sum fuzzy sets for Experiment II

MOTER CURRENT I

I universe \ fuzzy sets	ZE	S	M	B
1.	1.	0.	0.	0.
1.37	0.75	0.25	0.	0.
1.75	0.5	0.5	0.	0.
2.12	0.25	0.75	0.	0.
2.5	0.	1.	0.	0.
2.87	0.	0.75	0.25	0.
3.25	0.	0.5	0.5	0.
3.62	0.	0.25	0.75	0.
4.	0.	0.	1.	0.
4.37	0.	0.	0.75	0.25
4.75	0.	0.	0.5	0.5
5.12	0.	0.	0.25	0.75
5.5	0.	0.	0.	1.
5.87	0.	0.	0.	0.75
6.25	0.	0.	0.	0.5
6.62	0.	0.	0.	0.25

Table 4.4 Motor current fuzzy sets for Experiment II

CONTROL VARIABLE U

U universe \ fuzzy sets	N	ZE	S	M	B	SB
50.	1.	0.	0.	0.	0.	0.
53.75	0.75	0.25	0.	0.	0.	0.
57.5	0.5	0.5	0.	0.	0.	0.
61.25	0.25	0.75	0.	0.	0.	0.
65.	0.	1.	0.	0.	0.	0.
68.75	0.	0.75	0.25	0.	0.	0.
72.5	0.	0.5	0.5	0.	0.	0.
76.25	0.	0.25	0.75	0.	0.	0.
80.	0.	0.	1.	0.	0.	0.
83.75	0.	0.	0.75	0.25	0.	0.
87.5	0.	0.	0.5	0.5	0.	0.
91.25	0.	0.	0.25	0.75	0.	0.
95.	0.	0.	0.	1.	0.	0.
98.75	0.	0.	0.	0.75	0.25	0.
102.5	0.	0.	0.	0.5	0.5	0.
106.25	0.	0.	0.	0.25	0.75	0.
110.	0.	0.	0.	0.	1.	0.
113.75	0.	0.	0.	0.	0.75	0.25
117.5	0.	0.	0.	0.	0.5	0.5
121.25	0.	0.	0.	0.	0.25	0.75
125.	0.	0.	0.	0.	0.	1.
128.75	0.	0.	0.	0.	0.	0.75
132.5	0.	0.	0.	0.	0.	0.5
136.25	0.	0.	0.	0.	0.	0.25

Table 4.5 Control variable fuzzy sets for Experiment II

Error - sum S \ U	Motor current I: 1.	1.37	1.75	2.12	2.5	2.87	3.25	3.62	4.0	4.37	4.75	5.12	5.50	5.87	6.25	6.62
-16.	93.13	93.13	93.13	93.13	93.13	93.13	93.13	93.13	93.13	93.13	93.13	93.13	93.13	93.13	93.13	93.13
-15.	93.13	93.13	93.13	93.13	93.13	93.13	93.13	93.13	93.13	93.13	93.13	93.13	93.13	93.13	93.13	93.13
-13.5	93.13	93.13	93.13	93.13	93.13	93.13	93.13	93.13	93.13	93.13	93.13	93.13	93.13	93.13	93.13	93.13
-12.	93.13	93.13	93.13	93.13	93.13	93.13	93.13	93.13	93.13	93.13	93.13	93.13	93.13	93.13	93.13	93.13
-10.5	93.13	93.13	93.13	93.13	93.13	93.13	93.13	93.13	93.13	93.13	93.13	93.13	93.13	93.13	93.13	93.13
-9.	93.13	93.13	93.13	93.13	93.13	93.13	93.13	93.13	93.13	93.13	93.13	93.13	93.13	93.13	93.13	93.13
-7.5	93.13	93.13	93.13	93.13	93.13	93.13	93.13	93.13	93.13	93.13	93.13	93.13	93.13	93.13	93.13	93.13
-6.	93.13	93.13	93.13	93.13	93.13	93.13	93.13	93.13	93.13	93.13	93.13	93.13	93.13	93.13	93.13	93.13
-4.5	93.13	93.13	93.13	93.13	93.13	93.13	93.13	93.13	93.13	93.13	93.13	93.13	93.13	93.13	93.13	93.13
-3.	93.13	93.13	93.13	93.13	93.13	93.13	93.13	93.13	93.13	93.13	93.13	93.13	93.13	93.13	93.13	93.13
-1.5	93.13	93.13	93.13	93.13	93.13	93.13	93.13	93.13	93.13	93.13	93.13	93.13	93.13	93.13	93.13	93.13
0.	93.13	93.13	93.13	93.13	93.13	93.13	93.13	93.13	93.13	93.13	93.13	93.13	93.13	93.13	93.13	93.13
1.5	93.13	93.13	93.13	93.13	93.13	93.13	93.13	93.13	93.13	93.13	93.13	93.13	93.13	93.13	93.13	93.13
3.	93.13	93.13	93.13	93.13	93.13	93.13	93.13	93.13	93.13	93.13	93.13	93.13	93.13	93.13	93.13	93.13
4.5	93.13	93.13	93.13	93.13	93.13	93.13	93.13	93.13	93.13	93.13	93.13	93.13	93.13	93.13	93.13	93.13
6.	93.13	93.13	93.13	93.13	93.13	93.13	93.13	93.13	93.13	93.13	93.13	93.13	93.13	93.13	93.13	93.13
7.5	93.13	93.13	93.13	93.13	93.13	93.13	93.13	93.13	93.13	93.13	93.13	93.13	93.13	93.13	93.13	93.13
9.	93.13	93.13	93.13	93.13	93.13	93.13	93.13	93.13	93.13	93.13	93.13	93.13	93.13	93.13	93.13	93.13
10.5	93.13	93.13	93.13	93.13	93.13	93.13	93.13	93.13	93.13	93.13	93.13	93.13	93.13	93.13	93.13	93.13
12.	93.13	93.13	93.13	93.13	93.13	93.13	93.13	93.13	93.13	93.13	93.13	93.13	93.13	93.13	93.13	93.13
13.5	93.13	93.13	93.13	93.13	93.13	93.13	93.13	93.13	93.13	93.13	93.13	93.13	93.13	93.13	93.13	93.13
15.	93.13	93.13	93.13	93.13	93.13	93.13	93.13	93.13	93.13	93.13	93.13	93.13	93.13	93.13	93.13	93.13
16.	93.13	93.13	93.13	93.13	93.13	93.13	93.13	93.13	93.13	93.13	93.13	93.13	93.13	93.13	93.13	93.13

Table 4.6 Look-up table for Experiment IIa

Error - sum S	Motor current I															
U	1.	1.37	1.75	2.12	2.50	2.87	3.25	3.62	4.	4.37	4.75	5.12	5.5	5.87	6.25	6.62
-16.	93.13	93.13	93.13	93.13	93.13	90.63	87.50	82.31	65.	82.31	87.5	90.63	93.13	93.13	93.13	93.13
-15.	93.13	93.13	93.13	93.13	93.13	90.63	87.50	82.31	82.31	82.31	87.5	90.63	93.13	93.13	93.13	93.13
-13.5	93.13	90.63	87.50	87.50	87.50	87.5	87.50	89.58	89.58	89.58	89.58	90.63	93.13	93.13	93.13	93.13
-12.	93.13	90.63	87.50	82.31	65.	82.31	87.50	90.63	93.13	93.13	93.13	93.13	93.13	93.13	93.13	93.13
-10.5	93.13	93.13	93.13	93.13	93.13	93.13	93.13	93.13	93.13	93.13	93.13	93.13	93.13	93.13	93.13	93.13
-9.	93.13	94.62	96.5	99.62	110.	99.62	96.5	94.62	93.13	93.13	93.13	93.13	93.13	93.13	93.13	93.13
-7.5	93.13	94.62	96.5	96.50	96.5	96.5	96.5	94.62	93.13	93.13	93.13	93.13	93.13	93.13	93.13	93.13
-6.	93.13	94.62	96.5	99.62	110.	99.62	96.5	94.62	93.13	93.13	93.13	93.13	93.13	93.13	93.13	93.13
-4.5	93.13	94.62	96.5	96.50	96.5	96.5	96.5	96.5	96.5	96.5	96.5	94.62	93.13	93.13	93.13	93.13
-3.	93.13	93.13	93.13	93.13	93.13	94.62	96.5	99.62	110.	99.62	96.5	94.62	93.13	93.13	93.13	93.13
-1.5	93.13	88.25	86.25	91.15	93.13	94.62	96.5	96.5	96.5	96.5	96.5	94.62	93.13	93.13	93.13	93.13
0.	93.13	83.52	90.38	83.14	65.	82.31	87.5	90.63	93.13	93.29	93.5	93.85	95.	93.85	93.5	93.29
1.5	93.13	90.63	87.50	87.50	87.5	87.5	87.5	90.63	93.13	93.29	93.5	93.5	93.5	93.5	93.5	93.29
3.	93.13	90.63	87.50	82.31	65.	82.31	87.5	90.63	93.13	94.37	95.88	98.	102.5	82.5	95.88	94.37
4.5	93.13	91.70	90.38	90.38	90.38	90.38	90.38	88.49	87.5	89.55	93.13	93.13	93.13	93.13	93.13	93.13
6.	93.13	91.15	88.25	82.50	53.75	83.52	90.38	83.14	65.	83.14	90.38	83.52	53.75	82.5	88.25	91.15
7.5	93.13	91.15	88.25	88.25	88.25	88.96	90.38	88.49	87.5	89.55	93.13	93.13	93.13	93.13	93.13	93.13
9.	93.13	93.13	93.13	93.13	93.13	93.13	93.13	93.13	93.13	91.96	90.5	88.08	80.	88.08	90.50	91.96
10.5	93.13	93.13	93.13	93.13	93.13	91.15	88.25	88.25	88.25	90.24	93.13	91.45	90.5	90.5	90.50	91.96
12.	93.13	93.13	93.13	93.13	93.13	91.15	88.25	82.50	53.75	82.50	88.25	91.15	93.13	93.13	93.13	93.13
13.5	93.13	93.13	93.13	93.13	93.13	91.70	89.52	89.52	89.52	89.52	89.52	91.7	93.13	93.13	93.13	93.13
15.	93.13	93.13	93.13	93.13	93.13	90.63	87.5	82.31	82.31	82.31	87.5	90.63	93.13	93.13	93.13	93.13
16.	93.13	93.13	93.13	93.13	93.13	90.63	87.5	82.31	65.	82.31	87.5	90.63	93.13	93.13	93.13	93.13

Table 4.7 Look-up table for Experiment IIb

Error - sum S	U	Motor current I															
		1.	1.37	1.75	2.12	2.5	2.87	3.25	3.62	4.	4.37	4.75	5.12	5.5	5.87	6.25	6.62
	-16.	80.	80.	87.5	95.	95.	95.	95.	95.	0.	0.	0.	0.	0.	0.	0.	0.
	-15.	80.	80.	87.5	95.	95.	95.	95.	95.	0.	0.	0.	0.	0.	0.	0.	0.
	-13.5	65.	65.	87.5	110.	110.	110.	110.	102.5	0.	0.	0.	0.	0.	0.	0.	0.
	-12.	65.	65.	87.5	110.	110.	110.	110.	110.	0.	0.	0.	0.	0.	0.	0.	0.
	-10.5	61.25	61.25	81.56	110.	110.	110.	117.5	125.	125.	125.	125.	125.	0.	0.	0.	0.
	-9.	50.	51.87	88.91	110.	110.	110.	117.5	125.	125.	125.	125.	125.	0.	0.	0.	0.
	-7.5	61.25	61.25	81.56	110.	110.	110.	117.5	125.	125.	125.	125.	125.	0.	0.	0.	0.
	-6.	65.	65.	87.50	110.	110.	110.	117.5	110.	0.	0.	0.	0.	0.	0.	0.	0.
	-4.5	61.25	61.25	81.56	87.5	87.5	87.5	87.5	110.	110.	110.	110.	110.	0.	0.	0.	0.
	-3.	50.	51.87	61.25	65.	65.	65.	87.5	110.	110.	110.	110.	110.	0.	0.	0.	0.
	-1.5	53.75	53.75	61.25	65.	65.	65.	87.5	110.	110.	110.	110.	110.	0.	0.	0.	0.
	0.	50.	51.87	61.25	65.	65.	65.	65.	65.	0.	110.	110.	110.	110.	110.	110.	110.
	1.5	0.	65.	65.	65.	65.	65.	65.	65.	0.	102.5	102.5	102.5	102.5	102.	102.	102.
	3.	0.	65.	65.	65.	65.	65.	65.	65.	0.	95.	95.	95.	95.	95.	95.	95.
	4.5	53.75	53.75	61.25	61.25	61.25	61.25	61.25	65.	65.	65.	80.	80.	80.	80.	80.	80.
	6.	50.	51.87	53.75	51.87	50.	51.87	61.25	65.	65.	65.	65.	65.	65.	65.	65.	65.
	7.5	53.75	53.75	53.75	53.75	53.75	53.75	61.25	65.	65.	65.	72.5	72.5	72.5	72.5	75.	75.
	9.	0.	55.62	53.75	51.87	50.	51.87	53.75	55.62	0.	80.	80.	80.	80.	80.	80.	80.
	10.5	0.	55.62	53.75	53.75	53.75	53.75	61.25	65.	65.	65.	72.5	80.	80.	80.	80.	80.
	12.	0.	0.	0.	0.	0.	65.	65.	65.	65.	65.	72.5	80.	80.	80.	80.	80.
	13.5	0.	0.	0.	0.	0.	65.	65.	65.	65.	65.	72.5	80.	80.	80.	80.	80.
	15.	0.	0.	0.	0.	0.	65.	65.	65.	65.	65.	65.	65.	0.	0.	0.	0.
	16.	0.	0.	0.	0.	0.	65.	65.	65.	65.	65.	65.	65.	0.	0.	0.	0.

Table 4.8 Look-up table for Experiment IIIa

CONTROL VARIABLE U

U universe \ fuzzy sets	N	ZE	S	M	B	SB
50.	1.	0.	0.	0.	0.	0.
53.75	0.75	0.25	0.	0.	0.	0.
57.5	0.5	0.5	0.	0.	0.	0.
61.25	0.25	0.75	0.	0.	0.	0.
65.	0.	1.	0.	0.	0.	0.
68.75	0.	0.75	0.25	0.	0.	0.
72.5	0.	0.5	0.5	0.	0.	0.
76.25	0.	0.25	0.75	0.	0.	0.
80.	0.	0.	1.	0.	0.	0.
83.75	0.	0.	0.75	0.25	0.	0.
87.5	0.	0.	0.5	0.5	0.	0.
91.25	0.	0.	0.25	0.75	0.	0.
95.	0.	0.	0.	1.	0.	0.
98.75	0.	0.	0.	0.75	0.25	0.
102.5	0.	0.	0.	0.5	0.5	0.
106.25	0.	0.	0.	0.25	0.75	0.
110.	0.	0.	0.	0.	1.	0.
113.75	0.	0.	0.	0.	0.75	0.25
117.5	0.	0.	0.	0.	0.5	0.5
121.25	0.	0.	0.	0.	0.25	0.75
125.	0.	0.	0.	0.	0.	1.

Table 4.9 Control variable fuzzy sets for Experiment IV

MOTOR CURRENT I

I universe \ fuzzy sets	ZE	S	M	B
1.	1.	0.	0.	0.
1.37	0.75	0.25	0.	0.
1.75	0.5	0.5	0.	0.
2.12	0.25	0.75	0.	0.
2.5	0.	1.	0.	0.
2.87	0.	0.75	0.25	0.
3.25	0.	0.5	0.5	0.
3.62	0.	0.25	0.75	0.
4.	0.	0.	1.	0.
4.37	0.	0.	0.75	0.25
4.75	0.	0.	0.5	0.5
5.12	0.	0.	0.25	0.75
5.5	0.	0.	0.	1.
5.87	0.	0.	0.	1.
6.25	0.	0.	0.	1.

Table 4.10 Motor current fuzzy sets for Experiment IV

MOTOR VELOCITY Ω

ω universe \ fuzzy sets	ZE	S	M	B	SB
1000.	1.	0.	0.	0.	0.
1125.	0.75	0.25	0.	0.	0.
1250.	0.5	0.5	0.	0.	0.
1375.	0.25	0.75	0.	0.	0.
1500.	0.	1.	0.	0.	0.
1625.	0.	0.75	0.25	0.	0.
1750.	0.	0.5	0.5	0.	0.
1875.	0.	0.25	0.75	0.	0.
2000.	0.	0.	1.	0.	0.
2125.	0.	0.	0.75	0.25	0.
2250.	0.	0.	0.5	0.5	0.
2375.	0.	0.	0.25	0.75	0.
2500.	0.	0.	0.	1.	0.
2625.	0.	0.	0.	0.75	0.25
2750.	0.	0.	0.	0.5	0.5
2875.	0.	0.	0.	0.25	0.75
3000.	0.	0.	0.	0.	1.

Table 4.11 Motor velocity fuzzy sets for Experiment IV

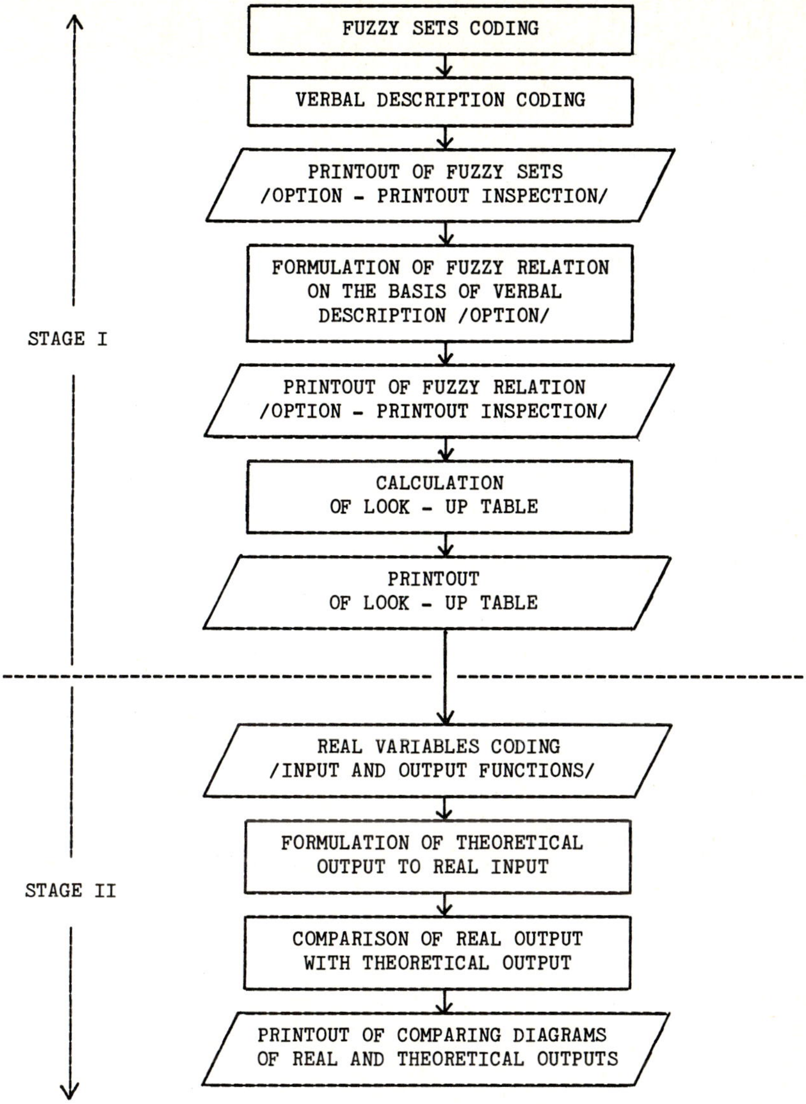

Fig. 4.1 Flow chart of experiment

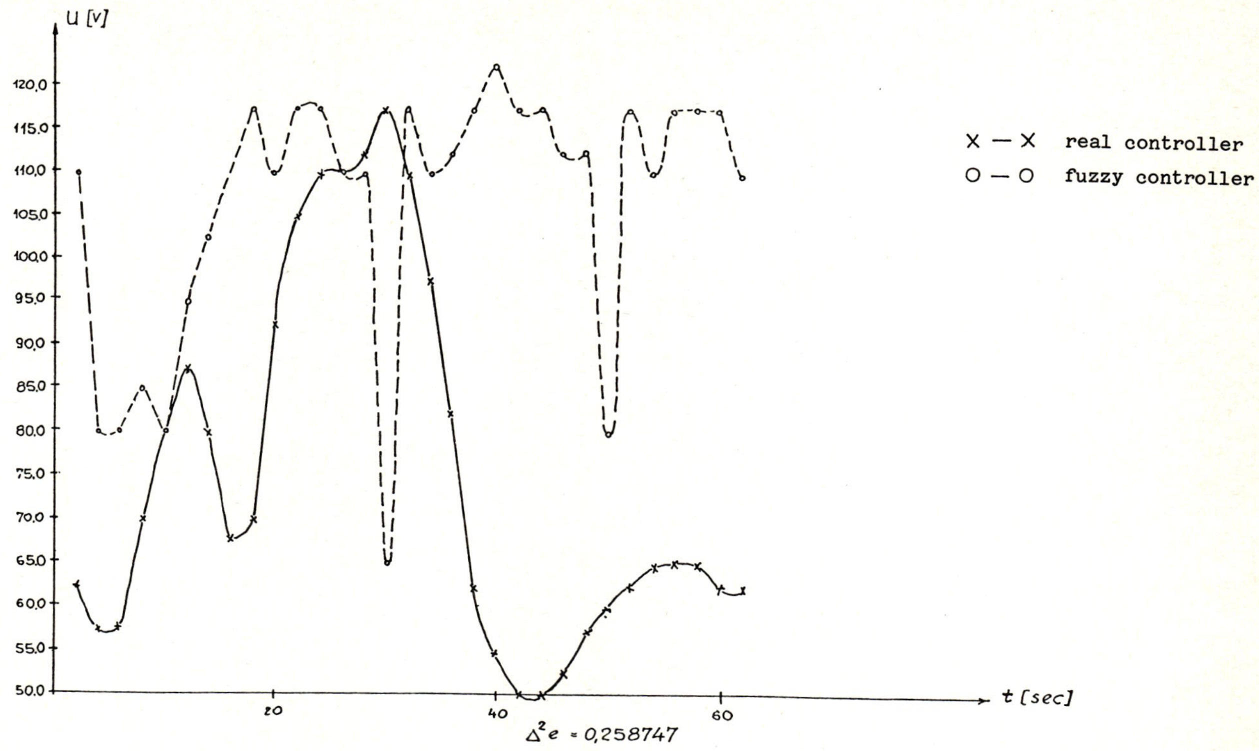

Fig. 4.2 Control variables for real and fuzzy controllers - Experiment Ia

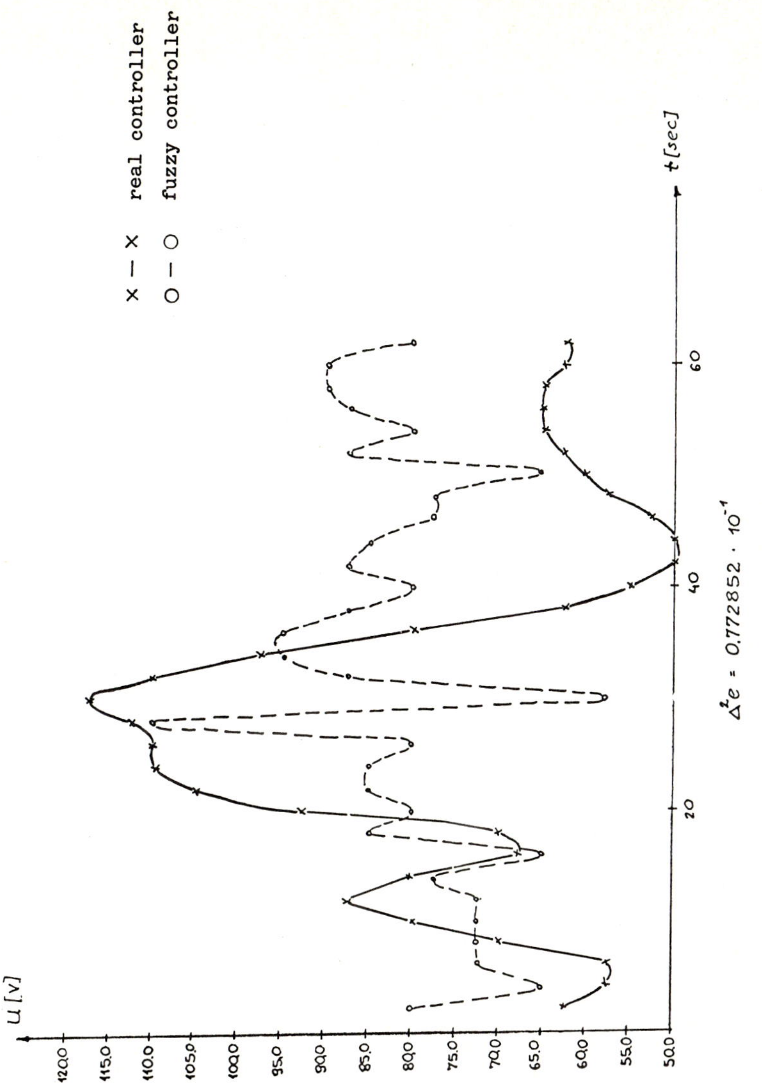

Fig. 4.3 Control variables for real and fuzzy controllers - Experiment Ib

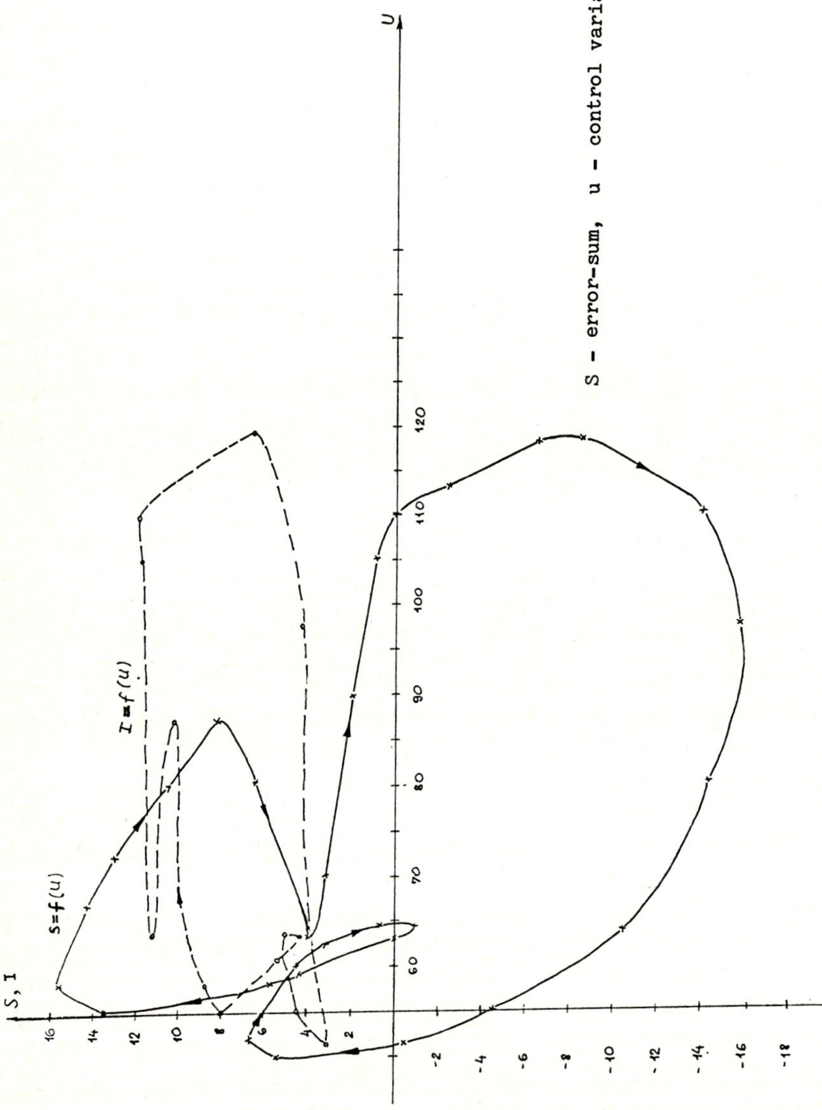

S - error-sum, u - control variable, I - motor current

Fig. 4.4 Dynamic phase characteristic of process control

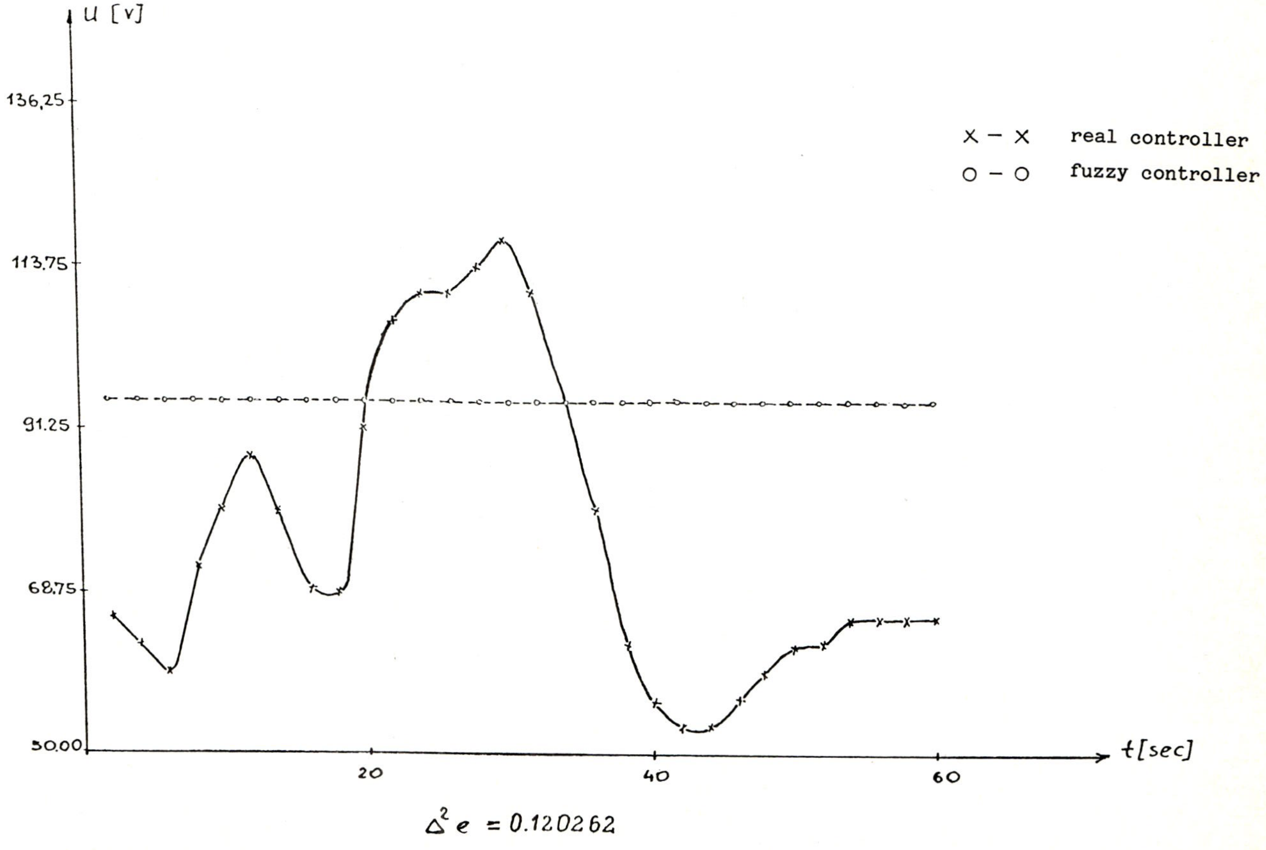

Fig. 4.5 Control variables for real and fuzzy controllers - Experiment IIa

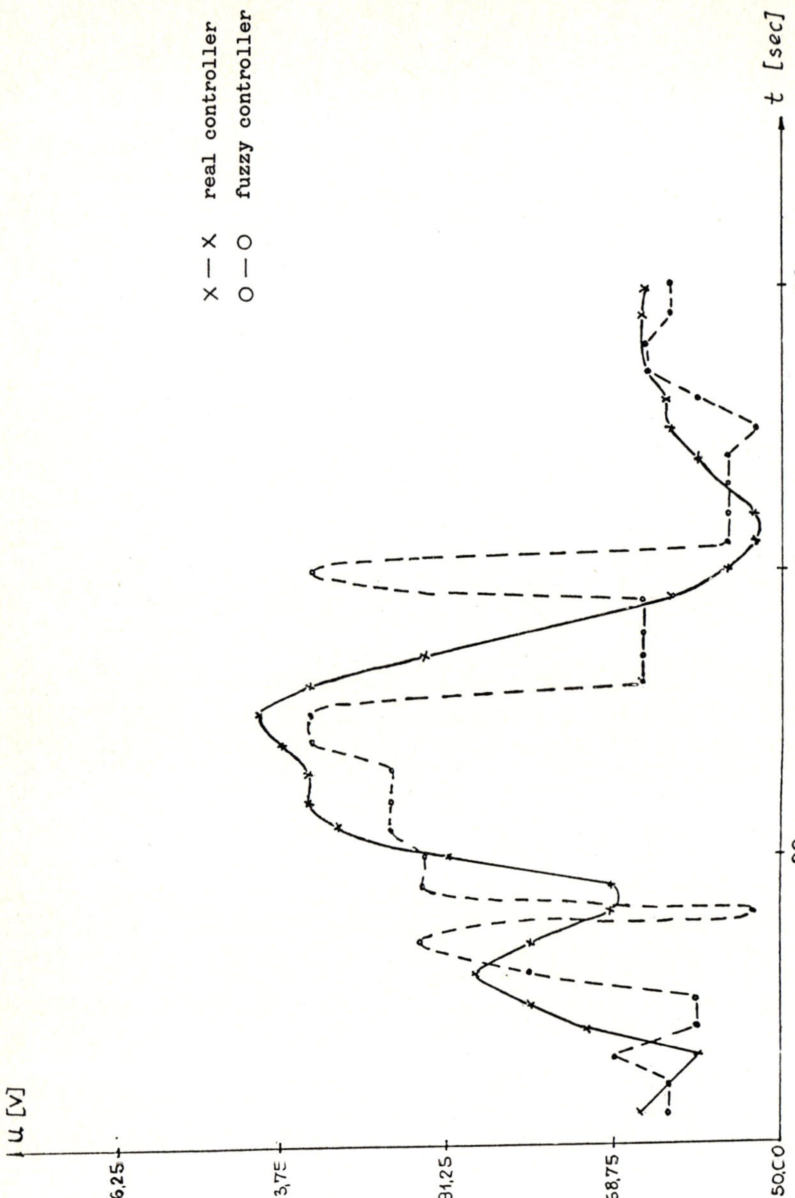

Fig. 4.6 Control variables for real and fuzzy controllers - Experiment IIb

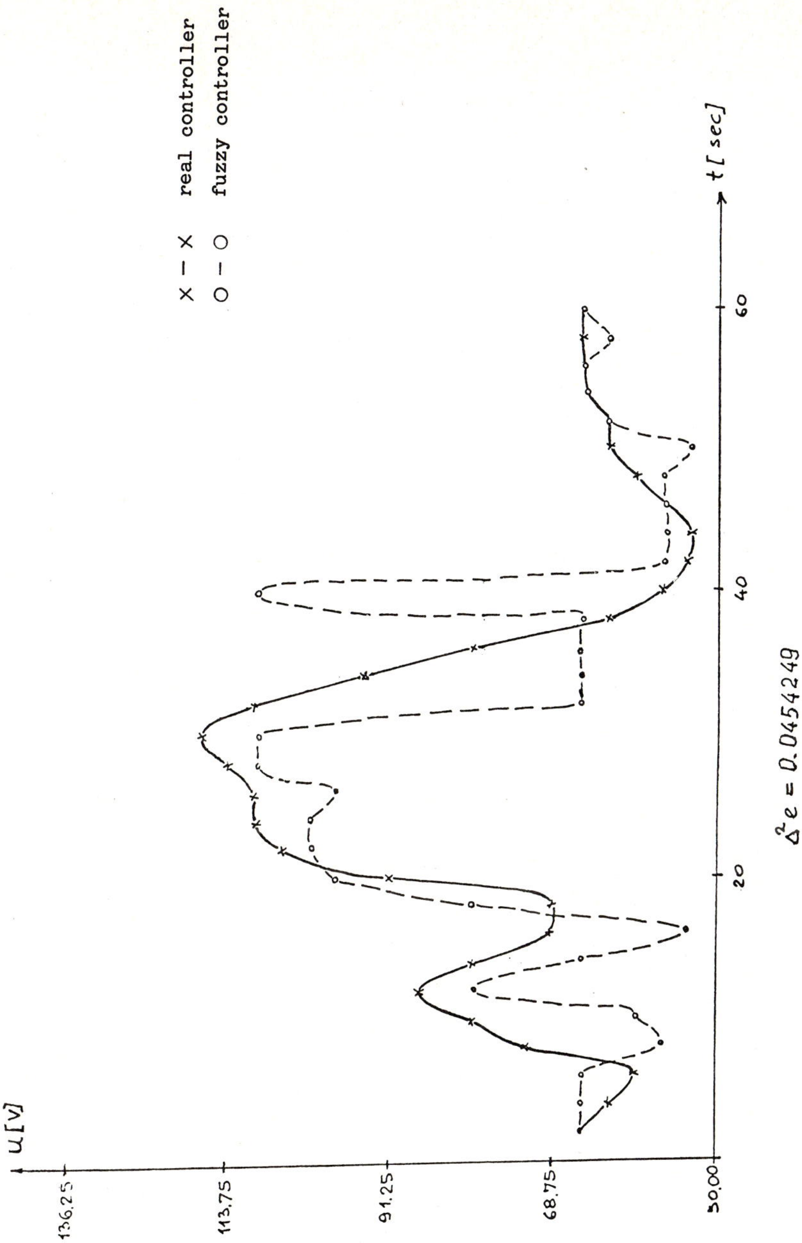

Fig. 4.7 Control variables for real and fuzzy controllers - Experiment IIc

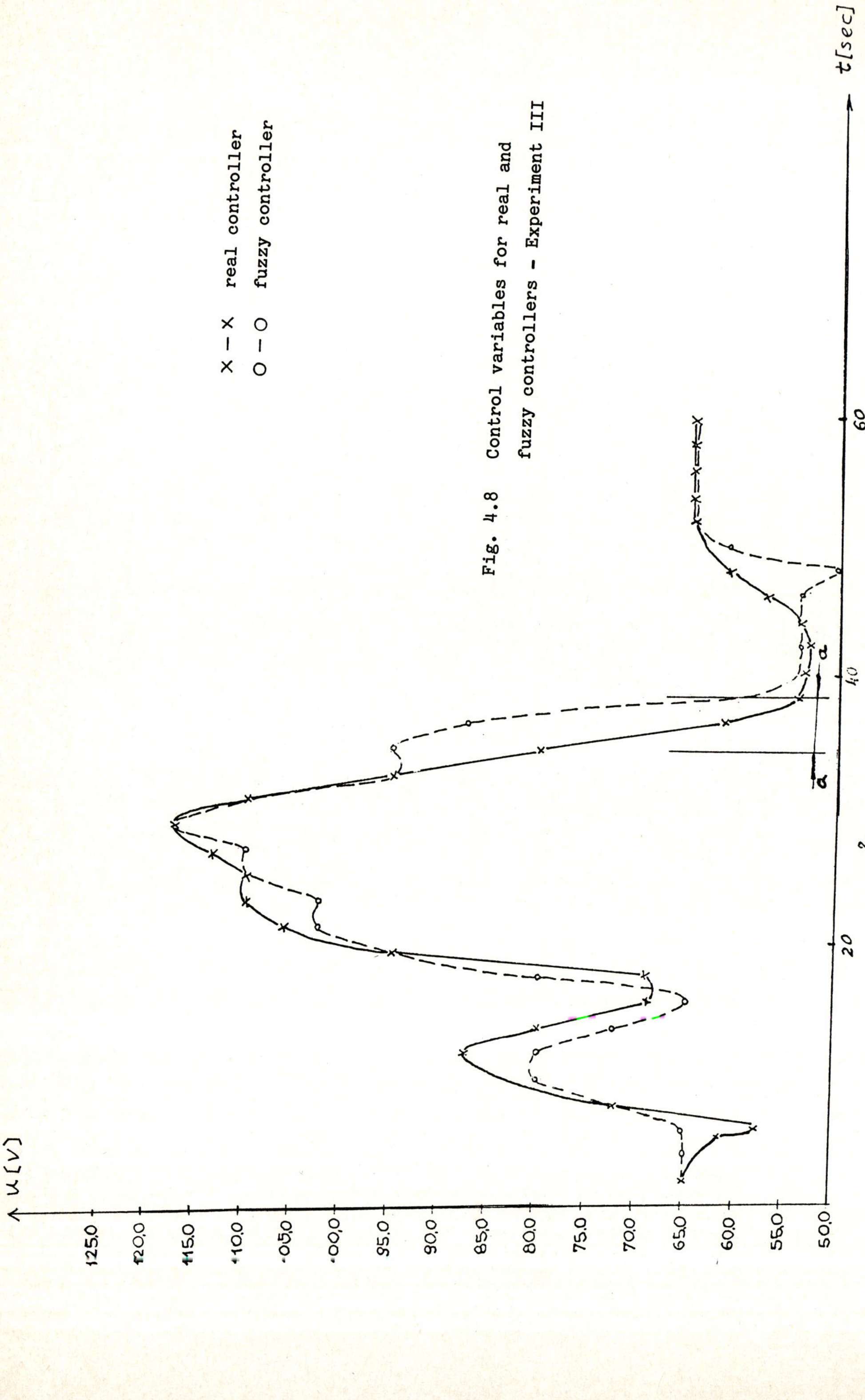

Fig. 4.8 Control variables for real and fuzzy controllers - Experiment III

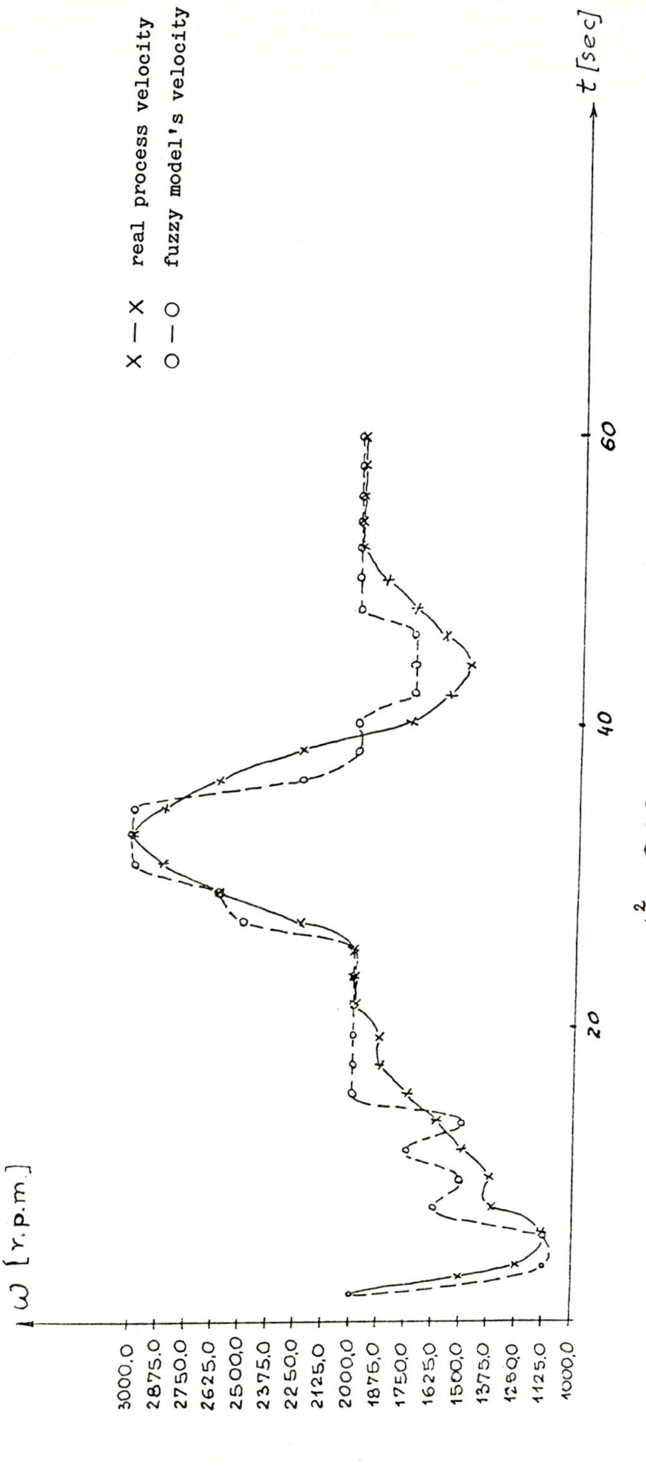

Fig. 4.9 Motor velocities for real process and fuzzy model of process - Experiment IV

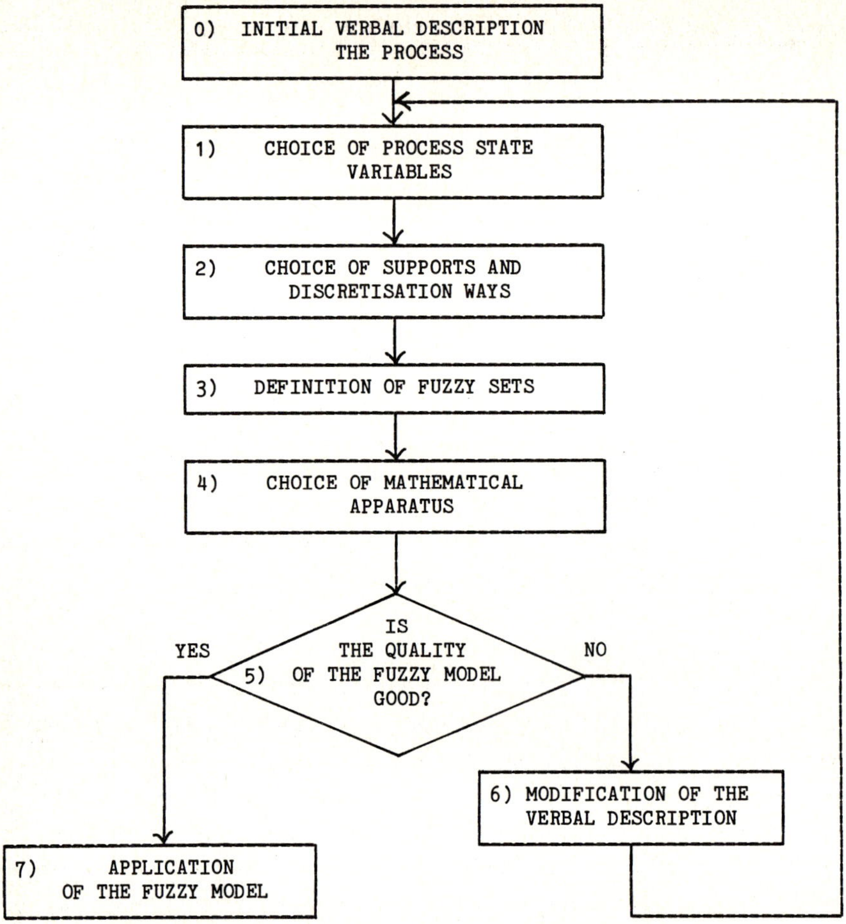

Fig. 5.1 Flow chart procedure of system designer

Industrial Applications of Fuzzy Control
M. Sugeno (ed.)

A MICROPROCESSOR BASED FUZZY CONTROLLER FOR INDUSTRIAL PURPOSES

Tsukasa Yamazaki* and Michio Sugeno**

* Systems and Control Design Department
Systems Engineering Division
JGC Corporation
1-14-1, Bessho, Minami-ku
Yokohama 232, JAPAN

** Department of Systems Science
Tokyo Institute of Technology
4259, Nagatsuta, Midori-ku
Yokohama 227, JAPAN

1. INTRODUCTION

Various fuzzy control algorithms have been proposed since Mamdani and Assilian [1] first applied fuzzy logic to the control of a pilot steam engine. Most of them are specialised to the control of specific prosesses and implemented by a rather versatile computer with adequate peripherals.

Because of a growing interest in fuzzy control among the process industries, it was thought that the construction of a microprocessor based fuzzy controller for general purposes would be a valuable step to explore the potential demands for fuzzy control. Designing a general purpose algorithm and implementing it with a microprocessor necessiate careful consideration of a new angle due to the required flexibility of control performance and the constraints from its hardware. As a first step, a prototype controller was developed for single input-single output control. The paper discusses some important results obtained from this study.

2. ALGORITHM

2.1 Inference Method

The basic fuzzy control algorithm adopted, including the inference method, is based on the control part of SOC (Self-Organising fuzzy Controller) investigated by Yamazaki [2]. The i-th fuzzy control rule is expressed with three fuzzy variables E (Error), CE (Change in Error), and CO (Change in output) in the following manner.

$$\text{If E is } E_i \text{ and CE is } CE_i \text{ then CO is } CO_i \qquad (2\text{-}1)$$

where, E_i, CE_i and CO_i belong to one of fuzzy labels of E, CE and CO respectively. They are normalized in common over the same discrete support set as shown in Fig. 1.

UNIVERSES OF DISCOURSE E, CE, CO

	-6	-5	-4	-3	-2	-1	0	1	2	3	4	5	6
NB	10	7	3	0	0	0	0	0	0	0	0	0	0
NM	3	7	10	7	3	0	0	0	0	0	0	0	0
NS	0	0	3	7	10	7	3	0	0	0	0	0	0
ZO	0	0	0	0	3	7	10	7	3	0	0	0	0
PS	0	0	0	0	0	0	3	7	10	7	3	0	0
PM	0	0	0	0	0	0	0	0	3	7	10	7	3
PB	0	0	0	0	0	0	0	0	0	0	3	7	10

NB : Negative Big
NM : Negative Medium
NS : Negative Small
ZO : Zero
PB : Positive Big
PM : Positive Medium
PS : Positive Small

Fig. 1 Fuzzy labels

Therefore parameters GE, GCE and GCO are introduced to map measured variables onto the values inside the algorithm, i. e., the support sets. This is a widely used technique and is very convenient for designing a general purpose controller because various processes can be controlled by just adjusting the parameters.

The fuzzy relation R_i of the conditional statement (2-1) is defined as the Cartesian product by multiplication instead of min operation, and the connective AND is also treated as the Cartesian product, which is interpreted as an interactive conjunction. The mathematical expression of the above is given below.

$$R_i \triangleq (E_i(e) \cdot CE_i(ce)) \cdot CO_i(co) \qquad (2\text{-}2)$$

$$e \in E, \quad ce \in CE, \quad co \in CO$$

The compositional rule of inference infers a control action CO_i^o from R_i when a control situation (E^o, CE^o) is given. With a simplification the fuzzy subsets of E^o and CE^o are treated as crisp sets, that is, E^o and CE^o have the membership value of 1.0 only at e^o and ce^o respectively and 0.0 elsewhere. CO^o inffered from all rules is given as follows.

$$CO^o(co) = \underset{i}{Max}\{E_i(e^o)\ CE_i(ce^o)\ CO_i(co)\} \qquad (2\text{-}3)$$

The use of the center of gravity procedure, as the interpretation of a fuzzy subset CO^o, gives the below in the membership from

$$co^o = \sum_{j=1}^{13} co_j \cdot \mu CO^o(co_j) \ / \ \sum_{j=1}^{13} \mu CO^o(co_j) \qquad (2\text{-}4)$$

The inference method abovementioned is quite simple but the usefullness of it has been shown in a simulation study [3].

2.2 Nest Arrangement of Control Rules

For a general use, especially industrial one, it is vital that a controller can produce non-abruptly rigorous or fine control actions corresponding to a control situation given. Though it is possible to meet the above requirement by using a large support set for CO, this leads to a huge increase of memory space or calculation time, or even the both whichever the support set is discrete or continuous. This is not desirable situation.

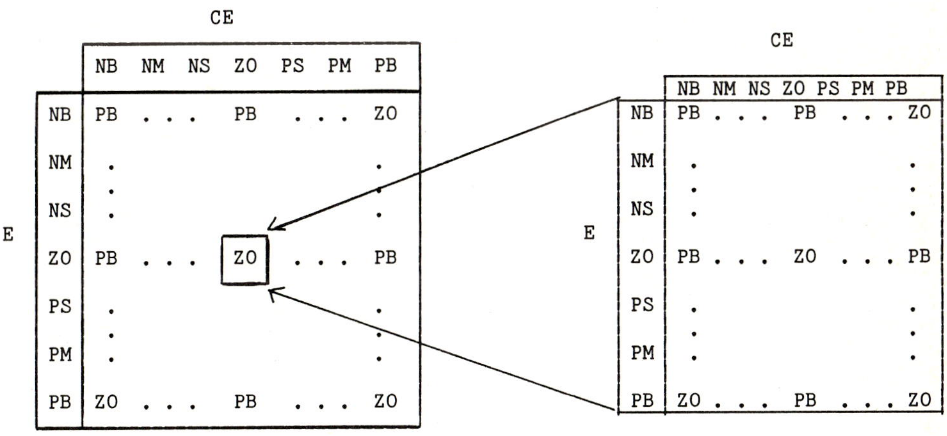

Fig. 2 Scheme of nesting control rules

For this reason, the control algorithm is arranged to have a "nested control window", in which the same set of control rules is used recursively and a gain for

CO is changed according to the degree of the nest. By adopting the control rule table representation, the procedure of nesting a set of control rules is illustrated in Fig. 2. When the control situation goes into the region specified on the table, for example, E = ZO and CE = ZO, the same set of control rules being used is mapped to this region together with the decrease of the controller gain GCO. This means that the control situation corresponding to the region is subdivided further and hence finer control output can be produced. Conversely, when the control situation goes out of the table, the set of control rules is mapped inversely accompanying the increase of the controller gain GCO accordingly. Thus, both the fine and rigorous control actions are achieved without the problems abovementioned.

The idea of "nested control window" is a natural extension of the human operator's behaviour. Because an operator tries first to bring the process value within a small deviation band around the setpoint, and then starts to near it at the setpoint when a large deviation of the process value occurs. In other words, operator's consciously or unconsciously try first to bring the process value to a roughly desirable situation and then to a precisely desirable one.

2.3 Selection of Operational Control Rules

With a view to easy operation and saving of memory space, the controller is designed so that appropriate control rules, as an operational set of rules, can be chosen from prospective control rules stored in a nonvolatile memory (PROM). The rule selection is specified by the rule numbers, which are assigned to the all prospective rules stored in the PROM. This is a practical method because the controller, like an ordinary industrial controller, has limited man-machine interface devices such as a digital display and a key board. It should be noted that the prospective control rules stored in the PROM were determined after a number of control simulations on various processes.

3. HARDWARE AND OPERATION

The hardware of the controller comprises a microcomputer board, an A/D and D/A converters board and an operational interface board as shown in Fig. 3.

The microcomputer board consists of a Z-80 microprocessor, two PROMs (2k byte), a RAM (1k byte) and auxiliary ICs. Both the A/D and D/A converters have a resolution of 8 bits. The operational interface board is mounted with two digital display sections, which indicate rule numbers and the other data separately, and with numerical and functional keys. Fig. 4 depicts the arrangements of such components on the board.

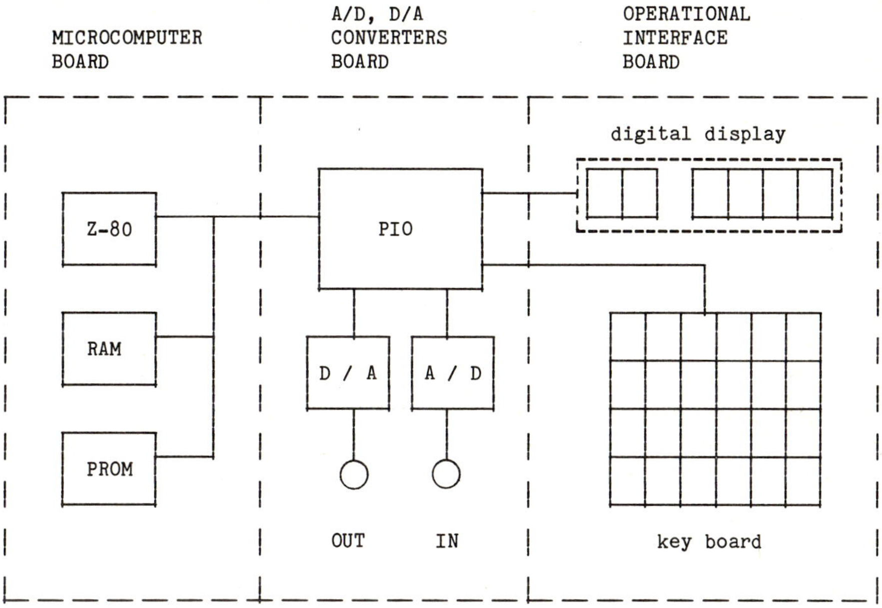

Fig. 3 Hardware configuration of prototype fuzzy controller

The controller needs several items to be input; controller parameters GE, GCE and GCO, operating parameters SP (setpoint) and TS (sampling period), and operational control rules. In the sequel, the operation of the controller is described by refering to Fig. 4.

(1) Setting and read-out of rules

The selection of operational control rules is initiated by the RIN key. Control rules can be stored in the RAM by repeating the procedure of pressing the ENT key after entering the rule number to be used until all the necessary rules are stored. The read-out of rule numbers stored in the RAM is started by the ROUT key, and then every time the ENT key is pressed, a rule number is shown sequentially on the rule number display section until all the stored rules are shown. The deletion of operational rules can be performed by entering the RDEL key, the rule number to be deleted and the ENT key in this sequence. All the operation for rule manipulation can be terminated by the END key.

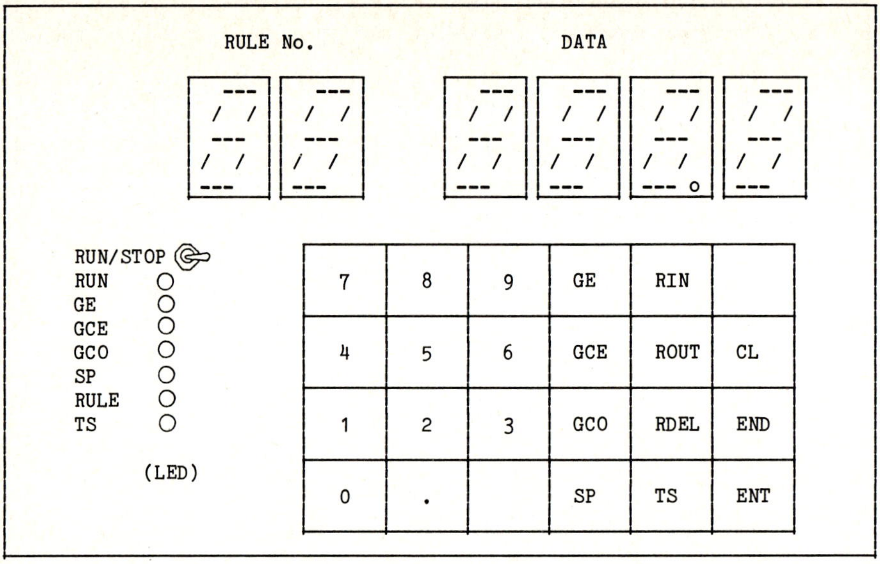

Fig. 4 Arrangement of operating interface board

(2) Setting and read-out of parameters

The operation for GE parameter is initiated by pressing the GE key, by which the value of GE stored in the RAM is shown on the data display section. When setting of GE is required, the CL key needs be pressed to clear the display instead of the END key. A value of GE can be stored in the RAM by the ENT key after entering the desired value with the numerical keys. The operation for other parameters GCE, GCO, TS and SP are performed likewise.

(3) Indicating function

Apart from the indication on the digital displays which is associated with the key operations abovementioned, the controller has two indication functions. One is the indication of a process value being controlled when the controller is at the RUN mode which is selected by a RUN / STOP switch. The other is the indication of the control rule in use, which is one of essential functions for a rule-based controller. On the rule number display section, it indicates the rule number which contributes most to the controller output inferred every sampling period. This function is a useful remedy for improving the operational control rule and make its procedure efficient.

4. PERFORMANCE OF THE CONTROLLER

A full investigation of the performance of the controller has not been done yet due to the shortage of time. The results obtained so far are described briefly here. To examine to what extent the controller can control difficult processes, an experiment was conducted by connecting it with an Apple-II microcomputer, in which a process model was programed, as shown in Fig. 5. The operational set of control rules is illustrated as a control rule table in Fig. 6. The transfer function of the process is given by $G(s)=e^{-Ls}/(1+Ts)$. Varying L and T, the performance of the controller was evaluated.

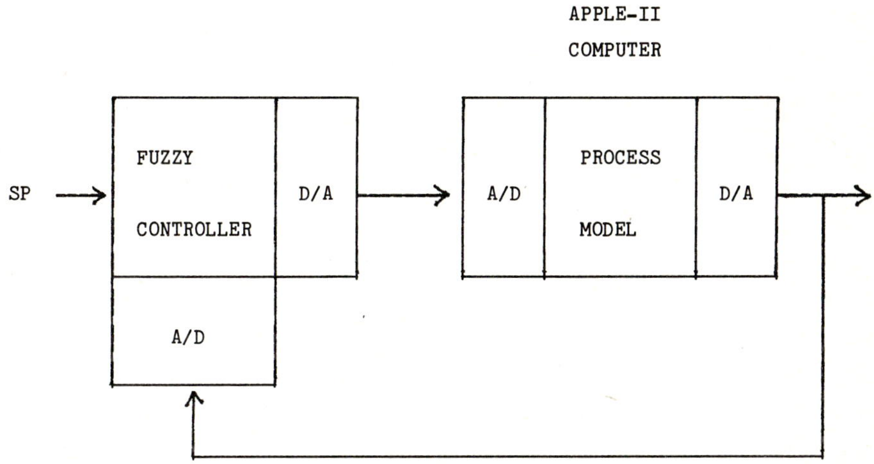

Fig. 5 Control flow of the experiment

CONTROL DECISION TABLE

		CHANGE IN ERROR (CE)						
		NB	NS	NS	ZO	PS	PM	PB
	NB	(--)**	(2)PB	(--)**	(8)PM	(10)PM	(13)PS	(--)**
E	NM	(16)PB	(--)**	(19)PM	(--)**	(22)PS	(--)**	(25)NS
R	NS	(--)**	(27)PM	(28)PS	(29)PS	(30)ZO	(32)NS	(34)NM
R	ZO	(35)PB	(--)**	(37)PS	(38)ZO	(39)NS	(--)**	(41)NB
O	PS	(42)PM	(44)PS	(46)ZO	(47)NS	(48)NS	(49)NM	(--)**
R	PM	(51)PS	(--)**	(54)NS	(--)**	(57)NM	(--)**	(60)NB
(E)	PB	(--)**	(63)NS	(66)NM	(68)NM	(--)**	(74)NB	(--)**

Fig. 6 Operational set of control rules used in the experiment

Fig. 7 shows one of the control responses which was obtained by a step change of the setpoint from 0 to 50% and then to 40%.

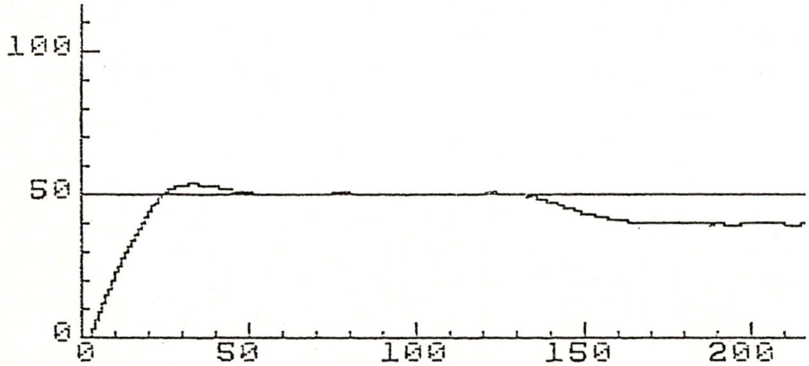

Fig. 7 Control response of the experiment

In view of the results, the performance of the controller is in general satisfactory as far as a ratio of L/T is less than 0.4. This fact indicates that this prototype controller compares to an optimally tuned PID controller in terms of ability to control difficult processes if one takes the ratio (L/T) as its index. This limitation is quite conceivable since the control rule form (2-1) does not include the element of time.

A human operator sometimes uses novel tactics, such as "Wait and See". Therefore, it is our belief that the function of variable interval control action should overcome the limitation found above. During the course of the experiment, two hardware problems were detected, though these are minor as can not seen from Fig. 7. One comes from the low resolution of A/D and D/A converters, and the other is noise into the analog circuit of the controller. However, these problems are not particular to this controller, and hence this point would not be discussed further just by stating that the improvement is under way.

5. CONCLUSION

The objective of this study is not to investigate a new fuzzy control algorithm but to demonstrate the possibility of constructing a microprocessor based fuzzy controller for industrial use. The study showed that the prototype fuzzy controller with the simple algorithm was quite capable, and that there were not many difficulties in constructing a microprocessor based fuzzy controller if careful considerations were given to the man-machine interface. It is hoped that the study reported here will encourage industrial people to explore the potential of fuzzy control.

6. ACKNOWLEDGEMENT

The authors are grateful to JGC corporation for the financial support which made this study possible. Many thanks are also given to Mr. Kawamoto and Mr. Tominaga for their assistance.

REFERENCES

[1] Mamdani, E. H. and Assilian, S.: An Experiment in Linguistic Synthesis with a Fuzzy Logic Controller, Int. J. of Man-Machine Studies, 7, 1-13, 1974

[2] Yamazaki, T.: An Improved Algorithm for a Self-Organising Controller, and Its Experimental Analysis, Ph.D Thesis, London University, 1982

[3] Yamazaki, T. and Mamdani, E. H.: On the Performance of a Rule-Based Self-Organising Controller; Proc. of IEEE Conf. on Applications of Adaptive and Multivariable Control, Hull, U.K. 50-55, 1982

Industrial Applications of Fuzzy Control
M. Sugeno (ed.)

THE APPLICATION OF FUZZY AND ARTIFICIAL INTELLIGENCE METHODS IN THE BUILDING OF A BLAST FURNACE SMELTING PROCESS MODEL

Hong Zhao* and Mou-chao Ma**

* Tianjin Institute of Industrial Automation and Instrumentation
Tianjin, China

** Institute of Psychology
Academia Sinica
China

In this paper the authors give a brief account of the building of an industrial process model by the fuzzy and AI methods. The Si (silicon) contained in the pig iron forecasting model of the blast furnace smelting process is built using general algorithms. Analysis of real data have proved that these algorithms are basically correct.

Key words : fuzzy system analysis, artificial intelligence, system identification, self-learning, process control

INTRODUCTION

The ability to identify, either directly from input-output data or from investigation of the process operator, a fuzzy model is an essential prerequisite for a practical theory of fuzzy control. It would also help give insight into the behaviour and basic structure of difficult processes, and in complex process generate the expert system. With this idea in mind we undertook a series of theoretical studies and have generated some application results in the set up of an expert system from blast furnace operators' experiences. The melting process of the blast furnace is an involved interaction process. Up to now, the blast furnace is the largest single piece of production equipment in industry. Owing to interval uneven distribution, some information difficulties in obtaining a given quantity, and some other stochastic disturbance factors, one is not able to get a applicable analytical model. So there is a barrier in using computers to control the smelting process in the furnace by the closed loop. In the two blast furnaces (about $1000M^3$) we use ALGOL and LISP to set up the semantic model networks of the knowledge base to forecast the desired silicon quantity at the pig iron by the practical measure systems in line on time. The anastomosis rate can become 85% (the error allowance : ±0.1%). This results in lowering the energy source consumption and leads to a larger economic significance.

1. The Simplest Fuzzy Semantic Model Network obtained by the Multi Variable Fuzzy Weighted Graph - Element Network Graph Simplified Method

An ordered triplet (V, A, $\underset{\sim}{\Phi}$) is called a fuzzily weighted graph. We write it as $\underset{\sim}{G}$. V is the vertex sets; A is the circular sets; $\underset{\sim}{\Phi} \subseteq \underset{\sim}{P}(R)$; $\underset{\sim}{\Phi}$ is the class of fuzzy subsets on R, the real number range, also the only full and simple map.

$$f : \underset{\sim}{A} \longrightarrow \underset{\sim}{\Phi}$$

The elements in Φ are called the fuzzy weights on a correspondence arc. $\underset{\sim}{\Phi}$ is called the fuzzy weighted class of $\underset{\sim}{G}$. If there is "if $\underset{\sim}{A}$ then $\underset{\sim}{B}$", let the two fuzzy variables $\underset{\sim}{A}$ and $\underset{\sim}{B}$ respectively correspond to the two vertex ordered couple in the fuzzy weighted graph (V_i, V_j) i, j = 1, 2, ... , N. Between the two vertices, fuzzy weights $\underset{\sim}{\Phi}_{i,j} \in \underset{\sim}{\Phi}$, also $\underset{\sim}{\Phi}_{i,j} = \underset{\sim}{R}_{i,j}$, as Fig. 1-1.

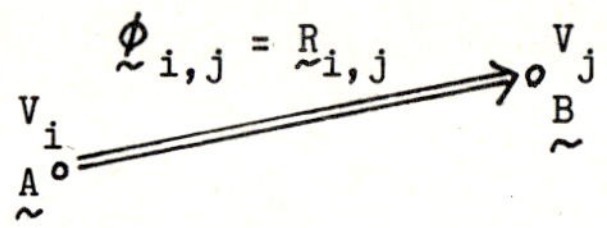

Fig. 1-1

$\underset{\sim}{R}_{i,j}$ is called the fuzzy relation between fuzzy variables $\underset{\sim}{A}$, $\underset{\sim}{B}$. According to the correspondence between the fuzzy weighted graph and the conditional statement given by Fig. 1-1, we can obtain an explanatory frame graph corresponding to the formula if $\underset{\sim}{A}_{t-L_i}$ and $\underset{\sim}{B}_{t-K_i}$ then $\underset{\sim}{B}_t$ as follows Fig. 1-2.

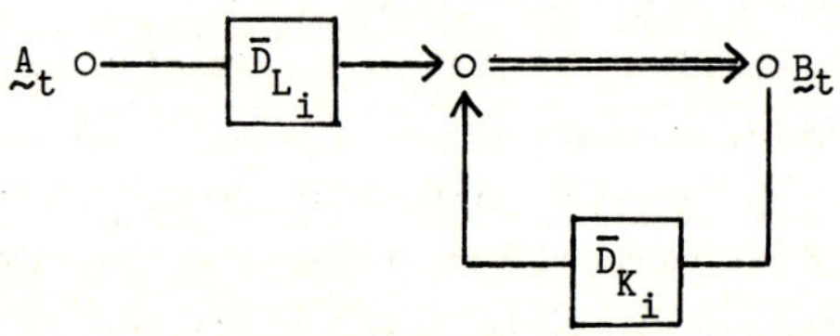

Fig. 1-2

Both $\bar{D}_{L_i}$ and $\bar{D}_{K_i}$ are the dead time operators of fuzzy variables. Just as $\underset{\sim}{A}_t$ through $\bar{D}_{L_i}$ acting becomes $\underset{\sim}{A}_{t-L_i}$, etc. From the point of view of the physical significance, we give "weights" to the time representation rule of the fuzzy relation as in shown Fig. 1-3.

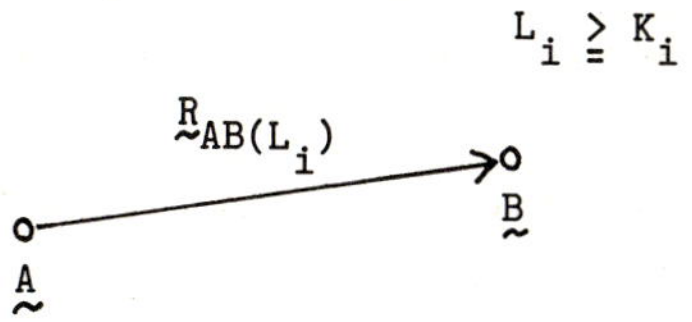

Fig. 1-3

At the same time, we set up the directed fuzzy element network graph corresponding to it, as shown in Fig. 1-4.

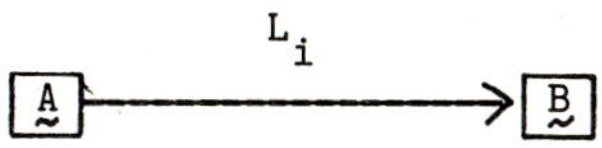

Fig. 1-4

We once attempted to use the method from [2] for finding the simplest gain of the graph to obtain the direct connection between the graph of input-output variables. But the gain equals the fuzzy relation matrix; and the fuzzy relation matrix is the semigroup algebraic structure without inverse elements. We have no way to quote the topological formula of the gain of the matrix signal flow in [2]; also there are many algorithms for fuzzy relation R, and it is difficult to select one, so instead of emphasizing the simplification of R, we try to avoid these difficulties by introducing the method of simplifying the fuzzy element network in order to solve the problem [3]. We have given an example of two inputs and one output as the sum-up for the discussion given above [4].

2. The Fuzzy Self-learning Correction Forecasting System

We have the following calculated solution, as shown in Fig. 2-1, from the simplified forecasting silicon semantic network.

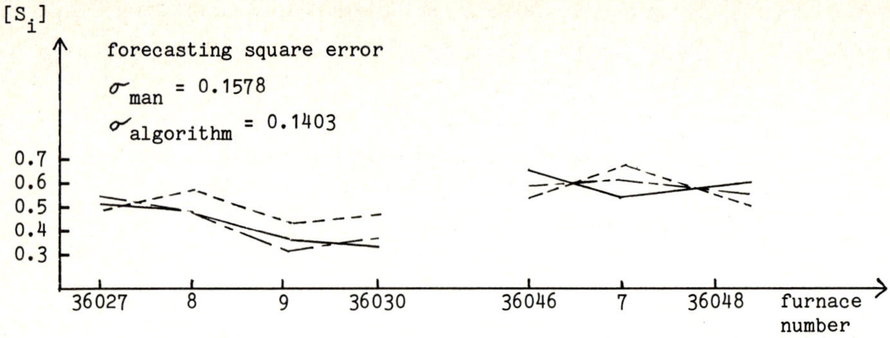

———	actual value
-------	human forecasting value
—-·—	algorithm forecasting value
$[S_i]$	the desired quantity of silicon at pig iron

Fig. 2-1

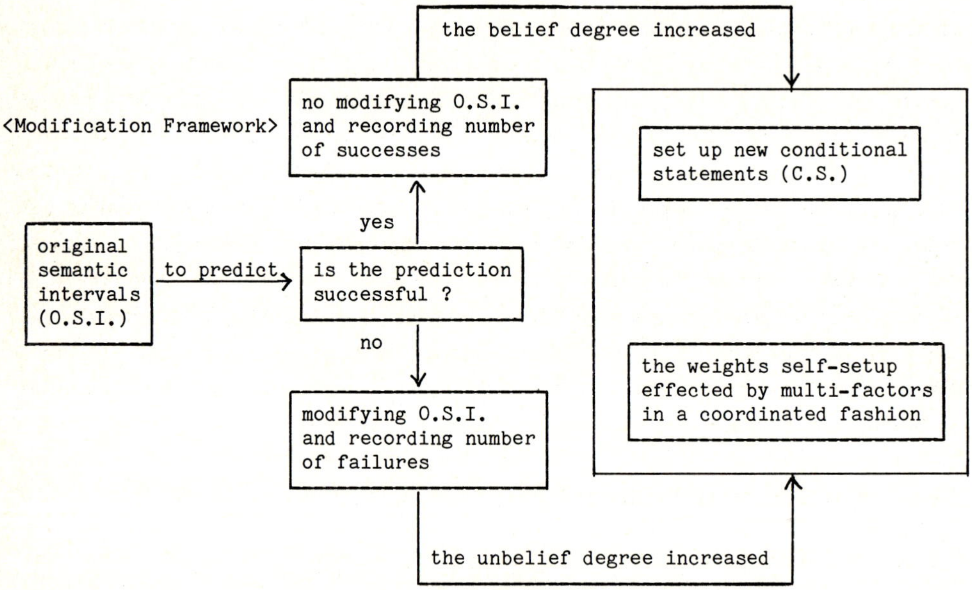

the modification of the O.S.I. ⟶ the modification of the C.S. ⟶ the modification of the rules in the expert system

Fig. 2-2

In the Fig. 2-1, the human forecasting values are mostly in accordance with the algorithm forecasting values in change trend, but unlike the actual values. The graph shows that in any fuzzy forecasting system, human experience should only be the primary form of the conditional statements (original map), the original maps may not conform with facts of the objective complex process, and also many maps are difficult to find; the system must be adjusted at random in practice. We summarize reference [5] as shown in Fig. 2-2.

A practical forecasting system should be a multi-factor and multi-link system. If the expected semantic of the latter link to the input is regarded as the practical semantic value of the former one, then the above mentioned self-learning correction algorithm can be recurred by using the decoupled algorithm from [4].

3. Increasing Learning Efficiency by the Memory Chunks Theory

As would be expected, the learning efficiency of a system in general increases as the learning time increases. But this efficiency reaches an asymptotic value at some time and does not increase thereafter as time increases. One of the reasons is probably due to the fact that in 'short-term memory' there is an 'information overload'. Can this be improved? In other words, is it possible to use more information than before, within its acceptable limitation? According to the study of memorization with information theory, the amount one can remember depends on the number of chunks in the stimulus to be remembered, rather than on the number of bits [6]. The former is regarded as conceptual units. It can be expected that the amount of information a system remembers can increase by reconstructing the chunks. For example, combining old information and forming others chunks within its acceptable limitation, that have a great deal of information associated with them [7]. Recently we attempted to put this idea into practice. In the fuzzy self-learning correction forecasting system reason to conceive of the process of reconstructing chunks as a moving window by exponent attenuation such as in Fig. 3-1.

Where, the width of the moving window $oo_1 = o_1o_2 = o_2o_3 = \dots = g(m)$, m - the number of theory chunks, implies some learning time after that the learning efficiency reaches an asymptotic value; determination of the parameter r of the curve is based on the result related to one's learning (reasoning) study conducted by [8] in which a curve of one's learning is characterized by an exponent function such as the following

$$y = \begin{cases} e^{r(x-1)} - 1, & x < i_0 \\ 1, & x \geq i_0 \end{cases}$$

let i_0 stand for a particular time at which a given rule is learned perfectly. Related to this, the parameter y is defined as the belief degree and 1-y as the unbelief degree; also let

$$y = \frac{\sum_{i=1}^{k} h \cdot y_i}{\sum_{i=1}^{k} y_i}$$

both formers are the practical belief degree and the unbelief degree respectively, where

$$h = \begin{cases} 1, & \text{using rule success} \\ 0, & \text{using rule failure} \end{cases}$$

X, K both stand for learning time, K = G(M), M - the number of practical chunks. These ideas have been employed to a self-learning forecasting system, and own results show increased learning efficiency.

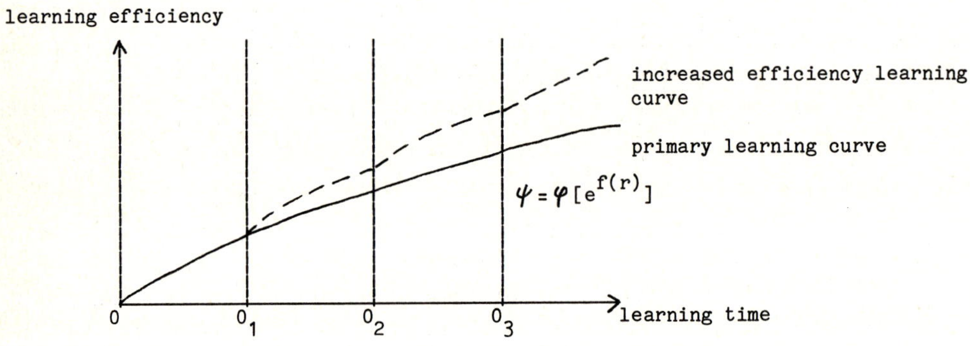

schematic diagram of the moving window by exponent attenuation

Fig. 3-1

REFERENCE

[1] Tong, R. M.: Synthesis of Fuzzy Models for Industrial Processes, Int. J. Gen. Systems, 4, 143-162, 1978

[2] Riegle, D. E. and Lin, P. M.: Matrix Signal Flow Graphs and an Optimum Topological Method for Evaluating Their Gains, IEEE Trans. on Circuit Theory, CT-19, 5, September 1972

[3] Kaufmann, A.: Introduction to the Theory of Fuzzy Subsets, 1, 235-242

[4] Zhao, Hong et al.: Multi Variable Fuzzy Weighted Graph — Element Network Graph and System Identification, the "Fuzzy Information Knowledge Representation and Decision Analysis" International Symposium preprint, 7, Marseilles, 1983

[5] Zhao, Hong and Li, Tai-hang: The Self-learning Correction Algorithm of Fuzzy System Identification, SINO-U.S. SYMPOSIUM on Fuzzy Methodology and Modern Analysis, 7, Beijing, 1984

[6] Coombs, C. H. et al.: Mathematical Psychology, 333-343, Mathesis Press, 1981

[7] Simon, H. A.: Models of Thought.

[8] Ma, Mou-chao, Li, Lai-fu and Hu, Xiao-dan: Reasoning Process and its Simulation, Personal Communication, 1983

Industrial Applications of Fuzzy Control
M. Sugeno (ed.)

AN ANNOTATED BIBLIOGRAPHY OF FUZZY CONTROL

Richard M. Tong

Advanced Information & Decision Systems
201 San Antonio Circle, Suite 286
Mountain View, CA 94040, USA.

This paper contains an annotated bibliography of papers on fuzzy control. The entries in the bibliography are primarily concerned with various aspects of the use of fuzzy sets in the design of controllers for dynamic systems. However, we have also included several references of a more general nature which seem to provide substantial insight on the fuzzy control problem. The bibliography contains annotated references to articles that have appeared between 1971 and 1983 inclusively in journals, conference proceedings, edited collections, and as technical reports. The bibliography proper is preceeded by an overview discussion of the main technical problems in fuzzy control system design.

INTRODUCTION

In compiling this bibliography we have attempted to construct a list of annotated references on *Fuzzy Control* that would be useful both to a novice and to someone who is already familiar with the main ideas in the field. The entries in the bibliography focus on papers that discuss various aspects of the use of fuzzy sets in the design of controllers for dynamic systems. We have also included several references of a more general nature which seem to provide substantial insight on the fuzzy control problem. The bibliography contains annotated references to articles that have appeared between 1971 and 1983 inclusively in journals, conference proceedings, edited collections, and as technical reports. In cases where work has appeared in substantially the same form in multiple sources, we have tended to include only the most recent and the most complete. Every attempt has been made to make the bibliographic information as accurate and complete as possible, but inevitably there will be some errors and omissions. We note, for example, that work on fuzzy control has appeared in Japanese, Polish, Russian and, most recently, Chinese. Unfortunately, little of this has been published in the West and the comments that follow apply only to English language publications.

The overview for the bibliography is divided into three parts. In the first section we present a brief historical commentary on the fuzzy control literature, pointing out some of its interesting features. In the second we highlight some of the key themes in the study of fuzzy control. (Of course, our own impressions do not necessarily coincide with that of others, and so the reader is referred to earlier reviews of the field for some insights into other issues that those authors have found to be significant [44],[69],[71].) In the third section, we construct a set of indexes that allow the user of the bibliography to find articles on the basis of a variety of sub-topics. The overview is followed by the bibliography proper.

SOME HISTORICAL COMMENTS

Let us start by considering Figure 1, which shows how work on fuzzy control has grown in scope, both in terms of the subject matter and the number of people involved, since the early beginnings. From the figure we see that the original idea of applying fuzzy sets to control system problems is due to Zadeh [122], [123]. Indeed if we read these early papers carefully, we can see that many of the key ideas in fuzzy control are already in evidence. Perhaps this is not surprising given his significant contributions to control theory.

Despite those very early contributions however, the catalyst for a large amount of work in fuzzy systems seems to have been the efforts of Mamdani and his colleagues [62], [74] at Queen Mary College, London. For example, the work of Kickert [60], Rutherford [90] and Ostergaard [83] is all based on the ideas generated by Mamdani's group, albeit with minor variations to suit the particular application. However, most of these applications have remained in the laboratory with the notable exception of Ostergaard's work which led to the development of a commercially available controller for cement kilns [49]. Proposals for applying the techniques to more complex processes have appeared (e.g., Tong, *et al.* [106]) but we are not aware of any systems that go beyond the research prototype stage.

As awareness of this novel approach to process control spread, there were an increasing number of attempts to analyse and formalise the concept of a fuzzy controller. At first these efforts were restricted to analyses that focussed on the particular structure imposed by the the rule-based paradigm, but more mature views have subsequently appeared (e.g., Braae and Rutherford [10], Burdzy and Kiszka [14], Kania, *et al.* [56]). Indeed, it seems that a large amount of effort in recent years has been devoted to defining the concept of a *fuzzy dynamic system* and analysing its behaviour as a precursor to developing some kind of control problem and its solution. While some of this work is directly applicable to the the original fuzzy controller concept (e.g., Baldwin and Guild [5], Cumani [18], Tong [105]) much is rather abstract (e.g., Kacprzyk [53], Kandel [55]) and should properly be considered a contribution to mathematics rather than control engineering.

To date there is a substantial literature within the field of fuzzy set theory that deals with problems of dynamical systems and control. This bibliography contains 123 entries, and although this is a relatively small number compared to the current fuzzy set literature (for example, a recent search of the IEE's INSPEC database revealed 1406 entries with the word fuzzy and its derivatives somewhere in the search fields), it represents a clearly defined subset of the fuzzy enterprise. To give some impression of the growth and relative emphasis of the literature, we show in Figure 2 a histogram for the number of papers published by year, with the total divided into two categories: "applications" and "theory". This division is somewhat fuzzy, of course, but by application study we mean one that concentrates on a particular "real-world" control problem rather than some general class of systems.

Several things are apparent from this histogram. Most obviously, after a slow build-up in the early seventies, there has been a steady output of papers in the last few years. Furthermore, it seems that the character of the effort has changed from being a practically oriented experimental undertaking into a more formal mathematical investigation. To some extent this reflects the changes in personnel within the field, but it also shows that the subject has come under much mathematical scrutiny.

As a final comment we note that the bulk of the fuzzy control work continues to originate in Europe (we include Soviet bloc countries) and Japan, with hardly any publications from US based authors. Indeed, given that fuzzy logic controllers draw extensively on the work of Lukasiewicz, there is a pleasing circularity in the fact that one of the most active groups working on the development of a fuzzy control theory is in Poland.

KEY THEMES IN FUZZY CONTROLLER STUDIES

We shall organize our discussion of themes under the two major headings of ***design*** and ***analysis*** that reflect the range of interests of those who have worked on fuzzy systems. Under each heading we shall mention some of the sub-themes that appear.

Themes in Design.

Research on controller design, as typified by Mamdani's approach, is primarily application oriented. It is concerned with devising a methodology that will allow the control engineer to develop a specification for a controller and then implement it. To illustrate some of the issues involved consider the functional diagram of a fuzzy control system shown in Figure 3. The fuzzy controller consists of five major elements: a rule-base of control statements, a database that contains the operational definitions of the fuzzy sets used in defining the control rules, a "condition" interface which transforms the non-fuzzy inputs it receives from the process into a fuzzy set form, an "action" interface which translates the fuzzy control action into a non-fuzzy one, and a computation unit in which the actual fuzzy control action is computed.

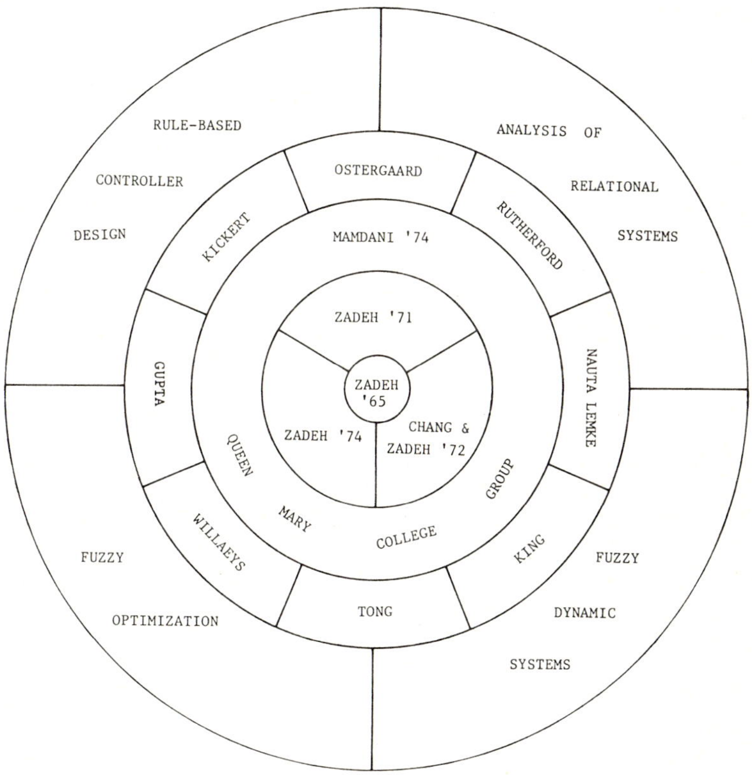

Figure 1 The Growth of Fuzzy Control

The main design problem is, of course, to assemble the set of control statements that constitute the rule-base. Since there are no systematic procedures for doing this, all reported applications have resorted to a trial-and-error approach. However, with increased experience some tentative design principles have been exposed. These have been mainly to do with developing a rule-base that is "complete" in some way (e.g., Braae and Rutherford [11], and Czogala and Pedrycz [22]). This concept is concerned with ensuring that the controller is always in a situation to be able to compute a meaningful control action given the current process conditions. That is, it embodies an intuitive notion that we should never be in a situation in which we do not know what to do. The concept can be defined in either the linguistic domain of the rules themselves or in the domain of the underlying physical process condition space.

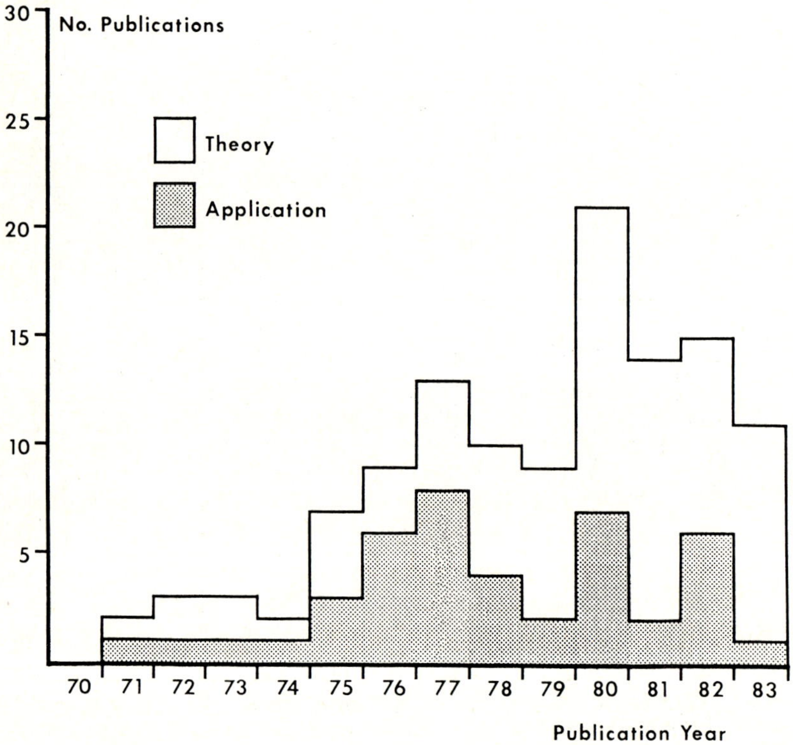

Figure 2 Histogram of Fuzzy Control Papers

Most rule-bases also assume that the condition statements will relate to process variables and their rates of change, and that the control actions will relate to control variables or their increments. Thus we see an attempt to use classical three-term control ideas in fuzzy controller design. This is entirely appropriate in the absence of a design theory, and shows how expert knowledge about process control can enter at a higher level than specification of condition-action pairs.

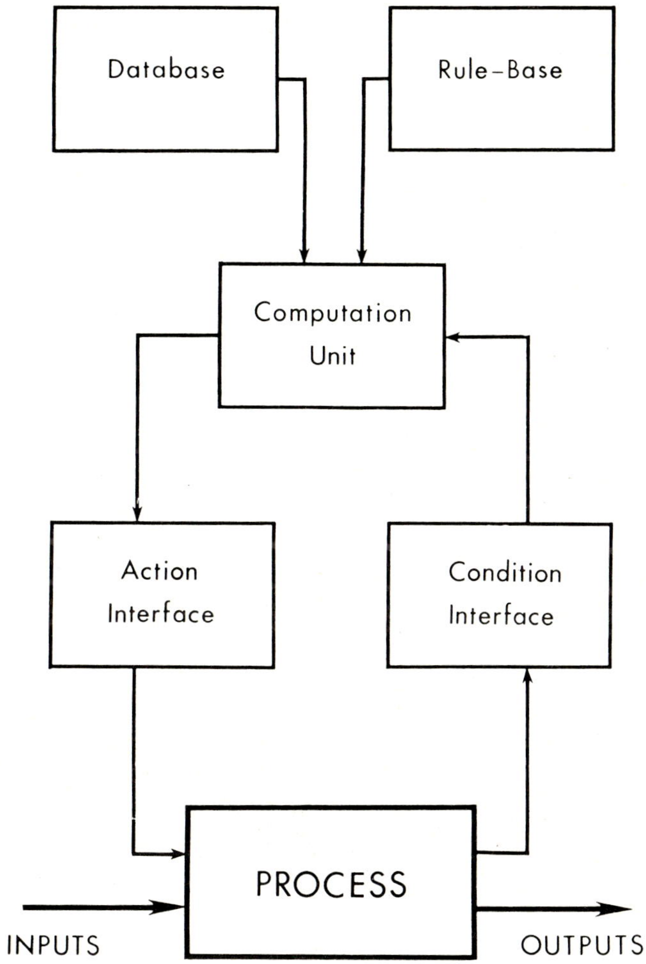

Figure 3 Functional Diagram of a Fuzzy Controller

Although construction of the rule-base is the crucial aspect of fuzzy controller design it is also the most difficult. Perhaps not surprisingly then, attempts have been made to automate this task. Among the most prominent is the work of Procyk and Mamdani [87] in which by invoking higher-level control knowledge about acceptable process response to particular kinds of disturbance, it is possible to "learn" a rule-base in its entirety. Another significant effort is that of Czogala and Pedrycz [25] in which control rules are generated via a performance index and the subsequent solution of a relational equation.

To make the fuzzy controller operational it is necessary to define the other elements in the system. Of those, the one which has received the most attention is the computational unit. In all existing controllers this involves adopting some underlying multiple-valued logical implication to translate the rules (most commonly either the outer product form first used by Mamdani, or that defined in Lukasiewicz's nondenumerably infinite logic), and using Zadeh's compositional rule of inference to compute the conclusion to what is essentially a generalized form of *modus ponens*. We find several papers that discuss the appropriateness of one implication over another (e.g., Baldwin and Pilsworth [6], and Bandler and Kohout [7]) and some commentary on the validity of the compositional rule (e.g., Mizumoto [78], and Sugeno and Takagi [94]).

Themes in Analysis.

The earliest attempts at analysis were primarily concerned with constructing appropriate theoretical representations, and with trying to develop some insights into meaningful measures of performance. Since most of the researchers had control theoretic backgrounds it was natural that they should be interested in extending such notions as state, controllability and stability, and with formulating equivalents to the output regulator and optimal control problems in their various guises (see Chang and Zadeh [15], and Jain [50] for example).

Two areas of investigation can be distinguished based on the point at which we draw the interface between the fuzzy controller and the process. Referring back to Figure 3, we see that if we place the input and output interfaces inside the fuzzy controller boundary then the controller is competely deterministic and looks like a rather sophisticated look-up table. If on the other hand, we consider the input and output interfaces as part of the process then the controller is truly fuzzy (i.e., it maps fuzzy sets onto fuzzy sets). The former viewpoint emphasises the discreteness of the controller, and we see discussions of its similarity to a multi-level relay and attempts to construct a form of quantized phase plane analysis (e.g., Braae and Rutherford [12], and Kickert and Mamdani [61]). The latter viewpoint is, we believe, much richer conceptually and some interesting results have been obtained by constructing relational models of the system and its controller (e.g., Gorzalczany, *et al.* [39], and Kiszka [65]).

Independently of these efforts, which were largely motivated by the practical work with fuzzy controllers, there have been several studies of a more abstract nature. Most have been directed towards the development of very general representations of fuzzy dynamic systems. So we see the work of Kacprzyk [53] on various forms of fuzzy mathematical programming, the work of Kloeden [66] on fuzzy topological approaches to dynamic systems and the work of deGlas [36] on a very general approach to fuzzy systems that includes a fuzzy stability theory.

Yet other analytic directions can be seen in the work of Czogala [20] on fuzzy probabilistic controllers, and are range of authors who have been concerned with the problems of building fuzzy models of dynamic processes (e.g., Kickert [59], Hirota and Pedrycz [47], Tong [103]).

BIBLIOGRAPHIC TABLES

As indicated above, the literature on fuzzy control and associated questions is rather large. In an attempt to impose some structure on the corpus we have devised a simple classification scheme as shown in Table 1. Obviously, these classification categories have fuzzy boundaries but they serve to partition the bibliography and help the user of the bibliography find references of interest. In Table 2 we give cross references to the bibliography on the basis of these categories. Notice that some references are highlighted (i.e., are in bold-face). These are papers that we have determined to be central to understanding the issues in fuzzy control, either because they contain key contributions or because they summarize fuzzy control work in a particularly coherent way.

Table 1 Classification Scheme for Fuzzy Control Papers

Category	Explanation
Applications	Studies which concern themselves with specific applications; either simulations, laboratory scale plant, or actual processes.
Relational Analysis	Studies which perform analysis on the assumption that the system is governed by a time-evolving fuzzy relational equation.
Optimization	Studies which extend conventional optimal methods by introducing fuzzy goals or constraints.
Dynamic Systems	Studies which attempt to develop more general of fuzzy dynamic system analysis than those considered under Relational Analysis.
Modelling	Studies of fuzzy models and modelling methods for dynamic systems.
Uncertain Reasoning	Studies of the impact of uncertainty in rule-based reasoning systems.
General	A catch-all category for general discussions of fuzzy control, including previous reviews and commentaries.

Table 2 Cross References from Classifications to Bibliography

Category	References
Applications	2, 8, 28, 31, 39, 48, **49**, **60**, 62, 63, 67, 70, 72, **74**, 77, 83, 84, 89, 91, 97, 98, **106**, 107, 110, 111, 113, 117, 120, 121.
Relational Analysis	5, 8, 10, 11, 12, 14, 16, 17, **18**, **22**, 23, 24, 25, 28, 39, 40, 41, 51, **56**, 57, 61, 64, 65, 72, 73, 85, 86, 88, 90, **96**, 98, 99, 101, 105, 106, 107, 109, 112, 113, 118, 119, **122**.
Optimization	3, 4, 13, 27, 30, 32, 38, 52, **53**, 68, 93, 115, 116.
Dynamic Systems	9, 15, 19, 20, 21, 26, **36**, 37, 47, 50, 54, 55, 61, 66, 75, 79, 80, 81, 87, 92, 95, **122**.
Modelling	**22**, 23, 24, 25, 35, 41, 47, 59, 86, **96**, 102, 103, 104, 110, 114, 118, 119.
Uncertain Reasoning	**6**, 7, 34, 78, 94.
General	1, 29, 33, 42, 43, 44, 45, 46, 58, 69, 71, 73, 76, **82**, **100**, 108, 123.

BIBLIOGRAPHY

[1] M.A. Aizerman, "Some Unsolved Problems in the Theory of Automatic Control and Fuzzy Proofs", *IEEE. Trans. Automatic Control*, AC-22(1):116-118, 1977.

A brief discussion of what approximate reasoning could be in the context of control systems. Introduces the idea of "correct within pathological cases" together with appropriate axioms and rules of inference.

[2] J. van Amerongen, H.R. van Nauta Lemke, J.C.T. van der Veen, "An Autopilot for Ships Designed with Fuzzy Sets", *Proc. 5th IFAC/IFIP Int. Conf. on Digital Computer Applications to Process Control*, The Hague, 1977.

The paper describes a simulated autopilot design that is based on a fuzzy model of human performance on the same task. The paper contains some comparison results with a conventional PID controller.

[3] K. Asai, S. Kitajima, "A Method for Optimizing Control of Multi-Modal Systems using Fuzzy Automata", *Inform. Sciences*, 3(4):343-353, 1971.

Addresses the steady state optimization problem by means of a fuzzy searching algorithm that adapts itself as a result of its performance. Contains the details of the results found in the following entry.

[4] K. Asai, S. Kitajima, "Optimizing Control using Fuzzy Automata", *Automatica*, 8(1):101-104, 1972.

A brief description of the use of an adaptive fuzzy automaton as a tool for the solution of a steady state optimization problem.

[5] J.F. Baldwin, N.C.F. Guild, "Modelling Controllers using Fuzzy Relations", *Kybernetes*, 9(3):223-229, 1980.

This paper suggests an alternative formulation for the rule-based controller based on Lukasiewicz logic. The argument is nicely presented and the paper shows how this formulation transforms the controller from a multi-level relay into a continuous function. The paper also contains a new approach for tuning such controllers.

[6] J.F. Baldwin, B.W. Pilsworth, "Axiomatic Approach to Implication for Approximate Reasoning using Fuzzy Logic", *Fuzzy Sets and Systems*, 3:193-219, 1980.

A detailed study of various forms of fuzzy implication rule. The paper contains a statement of the desirable properties of the implication and then tests various forms against these properties. A form based on Lukasiewicz's infinite valued logic appears best.

[7] W. Bandler, L.J. Kohout, "Semantics of Implication Operators and Fuzzy Relational Products", *Int. J. Man-Machine Studies*, 12:89-116, 1980.

This paper attempts an understanding of the various forms of multi-valued implication by using them in the analysis of clinical data. Several measures of implication are derived.

[8] G. Bartolini, G. Casalino, F. Davoli, M. Mastretta, R. Minciardi, E. Morten, "Development of Performance Adaptive Fuzzy Controllers with Application to Continuous Casting Plants", In: *Cybernetics and Systems Research*, Trappl R. (ed), North-Holland, 1982.

In this paper the author compares fuzzy controller algorithms with self-tuning algorithms and claims that they give similar performance results when the adaption is based on modification of the membership functions for the input-output variables.

[9] D. Benlachen, M. Lamotte, "A Fuzzy Automaton Synthesis Method", *Automatica*, 17(2):297-306, 1981.

This paper discusses a method for the identification of a fuzzy automaton which includes the determination of the state vector dimension, the evolution function and observation function. A simple example of the method for controller design is given.

[10] M. Braae, D.A. Rutherford, "Fuzzy Relations in a Control Setting", *Kybernetes*, 7(3):185-188, 1978.

A brief discussion of the use of fuzzy relations in controller analysis. It contains a useful discussion of the output selection problem, concluding that the "centre-of-mass" method is to be preferred.

[11] M. Braae, D.A. Rutherford, "Theoretical and Linguistic Aspects of the Fuzzy Logic Controller", *Automatica*, 15(5):553-577, 1979.

An extended analysis of fuzzy controllers. The paper shows how the approach based on multi-level relays is an "analytic dead end". The paper gives a design technique based on "inverting" a low-order linguistic model of the process.

[12] M. Braae, D.A. Rutherford, "Selection of Parameters for a Fuzzy Logic Controller", *Fuzzy Sets and Systems*, 2(3):185-199, 1979.

An interesting description of the design methodology, whose main idea is that "gains" should be adjusted first to give approximately the behavior desired. Only then, it is claimed, should the rules be tuned to "optimize" the response in the linguistic phase plane. Several examples of two-input one-output controllers are considered.

[13] G.S. Britov, L.K. Reznik, "Optimal Control of Linear Fuzzy Systems", *Automation and Remote Control*, 42(4.1):462-465, 1981.

The optimal control of a fuzzy system is considered on the basis of a fuzzy criterion of exponential type. An example of the method applied to an instrument test system is presented.

[14] K. Burdzy, J.B. Kiszka, "The Reproducibility Property of Fuzzy Control Systems" *Fuzzy Sets and Systems*, 9(2):161-177,1983.

An excellent attempt to analyze the relational form of the fuzzy controller paradigm for properties that are analogous to continuity of real functions.

[15] S.S.L. Chang, L.A. Zadeh, "On Fuzzy Mapping and Control", *IEEE Trans. Systems, Man and Cybern.*, SMC-2(1):30-34, 1972.

A discussion of the fuzzy control problems in terms of various mappings. The paper shows how under certain conditions a precise goal can be attained with fuzzy observations and control provided that the observations become sufficiently precise as the goal is approached.

[16] W-M. Cheng, S-C. Ren, C-F. Wu, T-H. Tsuei, "The Intersection of Fuzzy Subsets and the Robustness of Fuzzy Control", *Proc. IEEE. Int. Conf. on Cybernetics and Society*, Boston, October, 1980.

An analysis of the rule-based controller in terms of an "intersection value" defined for fuzzy sets.

[17] W-M. Cheng, S-C. Ren, C-F. Wu, T-H. Tsuei, "An Expression for Fuzzy Controller", In: *Fuzzy Information and Decision Processes*, Gupta M.M. and Sanchez E. (eds), North-Holland, Amsterdam, 1982.

A slightly shorter, but essentially similar, exposition to the previous paper.

[18] A. Cumani, "On a Possibilistic Approach to the Analysis of Fuzzy Feedback Systems", *IEEE Trans. Systems, Man and Cybern.*, SMC-12(3):417-421, 1982.

An excellent discussion of fuzzy feedback systems. The paper gives some new interpretations of fuzziness in systems and shows how a variety of control problems may be solved in this framework.

[19] E. Czogala, "An Idea of Generalized Fuzzy Probabilistic Control Algorithm", *BUSEFAL*, (13):275-294, 1982.

A description of a class of generalized fuzzy controllers based on the concept of probabilistic sets and the various operations that can be performed on them.

[20] E. Czogala, "A Generalized Concept of a Fuzzy Probabilistic Controller", *Fuzzy Sets and Systems*, 11(3):287-297, 1983.

This paper deals with some theoretical aspects of probabilistic fuzzy controllers. Some basic operations on probabilistic sets and their distribution functions are discussed. The paper includes some examples.

[21] E. Czogala, W. Pedrycz, "The Construction of Fuzzy-Probabilistic Controller", *BUSEFAL*, (9):153-160, 1977.

This paper introduces an extended notion of fuzzy controller in which the input and output data are considered as probabilistic sets. The more conventional notion of a fuzzy controller is show to be embedded in the extended form. Some details of how such controllers might be constructed are given.

[22] E. Czogala, W. Pedrycz, "Some Problems Concerning the Construction of Algorithms of Decision Making in Fuzzy Systems", *Int. J. Man-Machine Studies*, 15(2):210-211, 1981.

A discussion of some issues that arise in constructing rule-based fuzzy controllers. The concepts of completeness, interactivity and competitivity are introduced together with a discussion of their usefulness in controller design.

[23] E. Czogala, W. Pedrycz, "On Identification in Fuzzy Systems and its Applications in Control Problems", *Fuzzy Sets and Systems*, 6(1):73-83, 1981.

This paper gives a procedure for estimating the finite discrete system relation from a sequence of input-output fuzzy sets. Some measures of goodness for the model so produced are given, and methods for generating efficient test sequences are discussed.

[24] E. Czogala, W. Pedrycz, "Control Problems in Fuzzy Systems", *Fuzzy Sets and Systems*, 7(3):257-274, 1982.

This paper deals with a variety of control problems that arise in fuzzy relational systems. The possibility of simultaneous identification and control, control with constraints and a simple form of optimal control are considered.

[25] E. Czogala, W. Pedrycz, "Fuzzy Rule Generation for Fuzzy Control", *Cybernetics and Systems*, 13(3):275-294, 1982.

This paper deals with the design of control algorithms in fuzzy systems described by means of fuzzy relational systems. Several methods which consider the relative importance of the goals and constraints on the process are considered. The influence on the form of the controller relation is discussed in detail and illustrated by means of some numerical examples.

[26] E. Czogala, W. Pedrycz, "On the Concept of Fuzzy Probabilistic Controllers", *Fuzzy Sets and Systems*, 10(2):109-121, 1983.

This papers deals with a new approach to the construction of the so-called fuzzy probabilistic controller in which the input-output data are treated as probabilistic sets. The authors show that the notion of a fuzzy controller is embedded in the concept they discuss. A detailed analysis of theoretical properties of these controllers is followed by some simple examples.

[27] V. Dimitrov, W. Wechler, "Optimal Fuzzy Control of Humanistic Systems", *Proc. 6th IFAC World Congress*, Boston, 1975.

A somewhat vague discussion of how fuzzy sets may be used as the basis for devising control policies in social systems.

[28] V. Dimitrov, D. Driankov, A. Petrov, "Fuzzy Equations and their Applications to Water Pollution Control", *Proc. IFAC Symp. on Environmental Systems Planning, Design and Control*, Kyoto, Japan, August, 1977.

The paper uses fuzzy relational equation descriptions of a water quality control problem and shows how this static problem may be solved.

[29] D. Dubois, H. Prade, *Fuzzy Sets and Systems: Theory and Applications*, Academic Press, New York, 1980.

An extensive collation of results in all areas of fuzzy set theory. The book contains a chapter on dynamic systems that is especially relevant.

[30] M.M. Etschmaier, "Fuzzy Controls for Maintenance Scheduling in Transportation Systems", *Automatica*, 16(3):255-264, 1980.

A description of the use of fuzzy sets in the design of a computer-based tool for aiding maintenance scheduling.

[31] M.J. Flanagan, "On the Application of Approximate Reasoning to Control of the Activated Sludge Process", *Proc. Joint Automatic Control Conf.*, San Francisco, August, 1980.

A description of a proposed control system based the concepts of truth functional modification. The system is designed to be interactive with the human operator providing the truth values of various propositions about the process state.

[32] L.W. Fung, K.S. Fu, "Characterization of a Class of Fuzzy Optimal Control Problems", In: *Fuzzy Automata and Decision Processes*, Gupta M.M., Saridis G.N. and Gaines B.R. (eds), North Holland, 1977.

An example of a direct extension of the classical notion of an optimal control problem to include fuzzy goals and constraints.

[33] A. Fusaoka, H. Seki, K. Takahashi, "A Description and Reasoning of Plant Controllers in Temporal Logic", *Proc. 8th Int. Joint Conf. on Artificial Intelligence*, Karlsruhe (W. Germany), August, 1983.

An attempt to address the controller design problem by describing the plant dynamics via a (non-fuzzy) temporal logic. This allows concepts such as stability and controllability to be represented, and a graph theoretic model of the logic enables reasoning to be performed to generate controllers.

[34] B.R. Gaines, "Stochastic and Fuzzy Logics", *Electronics Letters*, 11(9):188-189, 1975.

This brief paper shows how it is possible to regard stochastic and fuzzy logics as being derived from two different constraints on an underlying probability logic. A comparison of a controller analyzed from both points of view shows there to be little difference in performance.

[35] B.R. Gaines, "Sequential Fuzzy System Identification", *Fuzzy Sets and Systems*, 2:15-24, 1979.

A description of an algorithmic procedure for deriving the structure of non-deterministic systems. The results assume an underlying fuzzy stochastic automata and demonstrate that sequential fuzzy data may modelled naturally in terms of "possibility" vectors.

[36] M. deGlas, "A Mathematical Theory of Fuzzy Systems", In: *Fuzzy Information and Decision Processes*, Gupta M.M. and Sanchez E. (eds), North-Holland, Amsterdam, 1982.

This paper proposes a theory of dynamical fuzzy systems. It shows that a fuzzy system can be represented by a flow describing the evolution of this fuzzy system in the space of fuzzy subsets of the state space. A discussion of stability in fuzzy systems is also presented.

[37] M. deGlas, "Theory of Fuzzy Systems", *Fuzzy Sets and Systems*, 10(1):65-77, 1983.

An essentially similar paper to the previous entry but with somewhat more emphasis on the topological properties of the system flow.

[38] B. Gluss, "Fuzzy Multi-Stage Decision Making, Fuzzy State and Terminal Regulators and their Relationship to Non-Fuzzy Quadratic State and Terminal Regulators", *Int. J. Control*, 17(1):177-192, 1973.

An exposition of the use of fuzzy cost functions for certain types of fuzzy control problem.

[39] M.B. Gorzalczany, J.B. Kiszka, M.S. Stachowicz, "Some Problems of Studying Adequacy of Fuzzy Models", In: *Fuzzy Set and Possibility Theory*, Yager R.R. (ed), Pergammon Press, New York, 1982.

This paper describes a fuzzy static model of a d.c. motor. The implementation and structure of the model are analysed in terms of the "good mapping" property, static stability and sensitivity. An extended numerical example is also given.

[40] S. Gottwald, E. Czogala, "On Interactivity in Fuzzy Controllers", *BUSEFAL*, (12):72-76, 1982/83.

A short paper which discusses the problem of overlapping (i.e., interacting) control rules. The result of this phenomenon is that even though the inputs to the controller may match a single rule's antecedent exactly, the output from the controller will not be identically equal to that rule's consequent. Some necessary and sufficient conditions are given for the controller to be non-interactive.

[41] R. Govind, "Synthesis of Fuzzy Controllers for Process Plants", *Proc. IEEE Int. Conf. on Cybernetics and Society*, Tokyo, November, 1978.

A systematic procedure for deriving control rules is described. The method is based on the idea of a process "cause-and-effect" graph from which a decision table can be derived which itself generates the fuzzy rules by a process of sequential elimination.

[42] M.M. Gupta, "Estimation and Control in a Fuzzy Environment", *Automatica*, 11:209-212, 1975.

A report of the round table discussion held at the 3rd IFAC Symposium on Identification and System Parameter Estimation.

[43] M.M. Gupta, P.N. Nikiforuk, "Feedback Control of Industrial Processes via Fuzzy Set Theory: Some Remarks", *Proc. Joint Automatic Control Conf.*, Pittsburgh, October, 1978.

In this paper a brief introduction to fuzzy concepts in the design of fuzzy control algorithms is given together with a quick survey of some of the work in the field.

[44] M.M. Gupta, "A Survey of Process Control Applications of Fuzzy Set Theory", *Proc. 18th IEEE Conf. on Decision and Control*, San Diego, January, 1979.

A brief survey of work being done on fuzzy controllers.

[45] M.M. Gupta, Y. Tsukamoto, "Fuzzy Logic Controllers - A Perspective", *Proc. Joint Automatic Control Conf.*, San Francisco, August, 1980.

A short review of studies in fuzzy logic control. Several problems, both theoretical and practical, are discussed and some possible resolutions offered.

[46] L.A. Gusev, I.M. Smirnova, "Fuzzy Sets Theory and Applications", *Automation and Remote Control*, (5):66-85, May 1973.

A review of the concepts of fuzzy sets in terms of languages, grammars, algorithms and automata. The paper mentions application in the area of systems control.

[47] K. Hirota, W. Pedrycz, "Analysis and Synthesis of Fuzzy Systems by the Use of Fuzzy Sets", *Fuzzy Sets and Systems*, 10(1):1-14, 1983.

This paper deals with an application of probabilistic sets in system theory. In particular, it addresses some identification problems in systems described by means of max-min fuzzy relational equations.

[48] L.P. Holmblad, J.J. Ostergaard, "Fuzzy Logic Control: Operator Experience Applied in Automatic Process Control", *ZKG International*, 34(3):127-133, 1981.

A description of the only fuzzy controller in commercial operation. This paper is oriented towards a general audience. (See the following entry for a more technical version of the paper).

[49] L.P. Holmblad, J.J. Ostergaard, "Control of a Cement Kiln by Fuzzy Logic", In: *Fuzzy Information and Decision Processes*, Gupta M.M. and Sanchez E. (eds), North-Holland, Amsterdam, 1982.

A complete and detailed description of the controller. The paper contains a good discussion of the process problems and the controller design, together with the computer implementation. Several charts giving a comparison with other control procedures are included.

[50] R. Jain, "Outline of an Approach for the Analysis of Fuzzy Systems", *Int. J. Control*, 23:627-640, 1976.

This paper describes an attempt to fuzzify the usual algebraic operators used in linear systems. In particular, the concept of a fuzzy transfer function is derived. Some practical issues of data explosion seem to limit the usefulness of this approach.

[51] J. Jantzen, "A Generalized Fuzzy Relation Matrix", *Int. J. Man-Machine Studies*, 13:413-421, 1980.

This paper describes a general and operational way of performing binary algebraic operations on fuzzy sets.

[52] J. Kacprzyk, "A Branch-and-Bound Algorithm for the Multi-Stage Control of a Fuzzy System in a Fuzzy Environment", *Kybernetes*, 8(2):139-147, 1979.

The problem of determining a maximizing decision in the multistage control of a fuzzy system in a fuzzy environment is considered. A branch and bound algorithm is proposed as a solution; two example are given.

[53] J. Kacprzyk, *Multistage Decision-Making Under Fuzziness*, Verlag TUV Rheinland, Koln, 1983.

This book is an extensive survey of a variety of fuzzy multistage decision problems. It represents a compendium of the results derived by the author and which have appeared in other places.

[54] A. Kandel, "On the Modelling of Uncertain Systems", *Proc. Int. Congress on Applied systems Research and Cybernetics*, Acapulco, December 1980.

This paper describes a set of tools called "the theory of typicality" that can be used for investigating dynamical systems which are governed by differential equations with fuzzy coefficients.

[55] A. Kandel, "On the Control and Evaluation of Uncertain Processes", *IEEE Trans. Automatic Control,* AC-25(6):1182-1187, 1980.

This paper investigates the applicability of fuzzy processes and fuzzy statistics to the problem of modelling fuzzy systems. The fuzzy systems considered are governed by fuzzy differential equations. The paper is similar in spirit to the previous entry.

[56] A.A. Kania, J.B. Kiszka, M.B. Gorzalczany, J.R. Maj, M.S. Stachowicz, "On Stability of Formal Fuzziness Systems", *Inform. Sciences,* 22:51-68, 1980.

A presentation of some formal results that describe the stability of fuzzy systems. Several definitions of stability are suggested, and necessary and sufficient conditions for systems to be stable are developed. The concept of a "good mapping" property is also introduced as being a desirable property of fuzzy systems.

[57] A.A. Kania, M. Stachowicz, "Robustness of Fuzzy Relation Operator", In: *Fuzzy Set and Possibility Theory,* Yager R.R. (ed), Pergammon Press, 1982.

A paper which describes some properties of the relation derived from sets of fuzzy rules.

[58] W.J.M. Kickert, *Fuzzy Theories on Decision-Making,* Martinus Nijhoff, Leiden, 1978.

A critical survey of the application of fuzzy set theory to modelling and decision-making. The second half of the book contains an examination of various results for dynamic systems.

[59] W.J.M. Kickert, "Towards and Analysis of Linguistic Modelling", *Fuzzy Sets and Systems,* 2:293-307, 1979.

A discussion of the possibility of modelling systems in linguistic terms. A major part of the paper is devoted to the analysis of dynamic behaviour. In particular, the paper discusses convergence and stability.

[60] W.J.M. Kickert, H.R. van Nauta Lemke, "The Application of Fuzzy Set Theory to Control a Warm Water Process", *Automatica,* 12(4):301-308, 1976.

A well written description of the fuzzy rule-based approach as applied to a simple water heating process. The algorithm developed is claimed to perform as well as an optimized PID controller although containing only a small number of rules.

[61] W.J.M. Kickert, E.H. Mamdani, "Analysis of a Fuzzy Logic Controller", *Fuzzy Sets and Systems,* 1(1):29-44, 1978.

An analysis of fuzzy controllers as multi-level relays using describing functions. A form of stability analysis is introduced. The approach appears to be limited to single-input single-output systems.

[62] P.J. King, E.H. Mamdani, "The Application of Fuzzy Control Systems to Industrial Processes", *Automatica,* 13(3):235-242, 1977.

An overview discussion of their work on control of a stirred tank reactor and a steam engine. The paper gives many practical details and performance results.

[63] R.E. King, "Fuzzy Logic Control of a Cement Kiln Precalciner Flash Furnace", *Proc. IEEE Conf. on Applications of Adaptive and Multivariable Control,* Hull (UK), 1982.

[64] J.B. Kiszka, M.S. Stachowicz, A.A. Kania, M.B. Gorzalczany, "Analysis and Synthesis of Fuzzy Control Systems", *Proc. 3rd IFAC/IFORS Conf. on Dynamic Modelling and Control of National Economies,* Warsaw, June, 1980.

The paper discusses the stability and sensitivity of fuzzy control systems. Two methods for constructing controllers are compared.

[65] J.B. Kiszka, *On the Convergence of Fuzzy Control Processes*, Dept. EECS, Univ. California at Berkeley, ERL Memorandum UCB/ERL-M81/43, 1981.

Using the relational equation approach to fuzzy system description, this report presents a variety of results that allow an analysis of the convergent behavior of such systems.

[66] P.E. Kloeden, "Fuzzy Dynamical Systems", *Fuzzy Sets and Systems*, 7:275-296, 1982.

A mathematical description of fuzzy systems that allows a discussion of the relationship between fuzzy, multi-valued and crisp systems. The paper also defines fuzzy trajectories and investigates some of their properties.

[67] M. Kokawa, "Heuristic Approach to Pump Operations using Multi-Valued Logic", In: *Fuzzy Information and Decision Processes*, Gupta M.M. and Sanchez E. (eds), North-Holland, Amsterdam, 1982.

A detailed description of a simple use of fuzzy control notions to help control a pumping station. The results of comparing an automated and human control system are given.

[68] S.V. Komolov, S.P. Makeev, G.P. Serov, I.F. Shaknov, "Optimal Control of a Finite Automaton with Fuzzy Constraints and a Fuzzy Target", *Cybernetics*, 15(6):805-810, 1979.

A discussion of multi-stage decision making in which the constraints and goals are time dependent fuzzy functions.

[69] P.M. Larsen, "Industrial Applications to Fuzzy Logic Control", *Int. J. Man-Machine Studies*, 12(1):3-10, 1980.

A review of a variety of fuzzy control studies at the Technical University of Denmark.

[70] E.H. Mamdani, "Applications of Fuzzy Algorithms for Control of Simple Dynamic Plant", *Proc. IEE*, 121(12):1585-1588, 1974.

The original announcement of the use of fuzzy rules for system control.

[71] E.H. Mamdani, "Advances in the Linguistic Synthesis of Fuzzy Controllers", *Int. J. Man-Machine Studies*, 8(6):669-678, 1976.

A short review of work on fuzzy controller synthesis. It contains a discussion of stability and gives a good selection of references.

[72] E.H. Mamdani, "Application of Fuzzy Logic to Approximate Reasoning using Linguistic Synthesis", *IEEE Trans. Computers*, C-26(12):1182-1191, 1977.

A description of the work with the steam engine together with some more general comments on the use of fuzzy logic.

[73] E.H. Mamdani, "A Fuzzy Rule-Based Method of Controlling Dynamic Processes", *Proc. 20th IEEE Conf. on Decision and Control*, San Diego, December 1981.

An interesting attempt to focus on the intuitive and heuristic aspects of controller synthesis. The paper argues for a re-assessment of the needs of practical control engineers and the design tools they use.

[74] E.H. Mamdani, S. Assilian, "An Experiment in Linguistic Synthesis with a Fuzzy Logic Controller", *Int. J. Man-Machine Studies*, 7(1):1-13, 1975.

This is perhaps the key paper in the fuzzy controller literature. It describes in detail the original work done at Queen Mary College with the laboratory steam engine.

[75] E.H. Mamdani, N. Baaklini, "Prescriptive Methods for Deriving Control Policy in a Fuzzy Logic Controller", *Electronics Letters*, 11(25/26):625-626, 1975.

A brief discussion of a mechanism for adaptive fuzzy control.

[76] E.H. Mamdani, T.J. Procyk, "Application of Fuzzy Logic to Controller Design Based on Linguistic Protocol", In: *Progress in Cybernetics and Systems Research: Vol. III*, Klir G.J. and Ricciardi L. (eds), Wiley, Chichester, England, 1978.

A tutorial style presentation of the work on the steam engine and the prescriptive controller.

[77] E.H. Mamdani, B.S. Sembi, "Process Control Using Fuzzy Logic", In: *Fuzzy Sets: Theory and Applications to Policy Analysis and Information Systems*, Wang P.P. and Chang S.K. (eds), Plenum, New York, 1980.

A description of fuzzy controllers together with a description of a fuzzy controller for a cement kiln.

[78] M. Mizumoto, "Note on the Arithmetic Rule by Zadeh for Fuzzy Conditional Inference", *Cybernetics and Systems*, 12:247-306, 1981.

A detailed discussion of two new composition operators than appear to be "better behaved" than max-min, and have some useful properties.

[79] C.V. Negoita, "On Fuzzy Systems", *Proc. 4th Int. Congress on Cybernetics and Systems*, Amsterdam, August 1978.

[80] C.V. Negoita, D.A. Ralescu, "Fuzzy Systems and Artificial Intelligence", *Kybernetes*, 3:173-178, 1974.

This paper gives a functional notation for fuzzy systems and defines the concepts of controllability and observability.

[81] C.V. Negoita, A.C. Stefanescu, "On the State Equation of Fuzzy Systems", *Kybernetes*, 4:213-214, 1975.

This paper gives a fuzzy set theoretic description of state equations for linear systems.

[82] C.V. Negoita, D.A. Ralescu, *Application of Fuzzy Sets to Systems Analysis*, Birkhauser Verlag, Basel, 1975.

A mathematical treatment of various aspects of fuzzy systems in their widest sense. The book contains some valuable remarks on the forms of analysis needed to develop a a fuzzy control theory.

[83] J.J. Ostergaard, "Fuzzy Logic Control of a Heat Exchanger Process", In: *Fuzzy Automata and Decision Processes*, Gupta M.M., Saridis G.N. and Gaines B.R. (eds), North Holland, 1977.

A description of one of the early applications of the fuzzy controller paradigm.

[84] C.P. Pappis, E.H. Mamdani, "A Fuzzy Logic Controller for a Traffic Junction", *IEEE Trans. Systems, Man and Cybern.*, SMC-7(10):707-717, 1977.

Another application of the fuzzy rule-based paradigm. The paper contains many practical details and comparisons with more conventional approaches to traffic signal control.

[85] W. Pedrycz, "An Approach to the Analysis of Fuzzy Systems", *Int. J. Control*, 34(3):403-421, 1981.

A paper which presents a variety of mathematical definitions for use in fuzzy systems. In particular, the author discusses prediction, sensitivity and stability. This is basically a relational approach to system analysis.

[86] W. Pedrycz, *Fuzzy Control and Fuzzy Systems*, Technical Report TR-82-14, Dept. Mathematics, Delft Univ. Technology, July 1982.

An extended collection of the author's results which deal with the theory and applications of fuzzy relational equations to fuzzy control and fuzzy systems.

[87] T.J. Procyk, E.H. Mamdani, "A Linguistic Self-Organizing Process Controller", *Automatica*, 15(1):15-30, 1979.

A detailed description of an adaptive controller that uses a linguistic description of the control strategy as a basis for learning. The method uses fuzzy set theory to operationalize the concepts and is applied to several examples of simple non-linear multi-variable processes.

[88] G.P. Rao, D.A. Rutherford, "Approximate Reconstruction of Mapping Functions from Linguistic Descriptions in Problems of Fuzzy Logic Applied to System Control", *Cybernetics and Systems*, 12(3):225-236, 1981.

An interesting paper that describes a method for taking rule-based descriptions of control and transforming them into approximate analytic functions.

[89] D.A. Rutherford, G.A. Carter, "A Heuristic Adaptive Controller for a Sinter Plant", *Proc. 2nd IFAC Symp. on Automation in Mining, Mineral and Metal Processing*, Johannesburg, RSA, September, 1976.

This paper describes the application of the fuzzy rule-based approach to control of permeability of the raw mix on a sinter process. It gives some practical details of the algorithm and comparisons with human control.

[90] D.A. Rutherford, G.C. Bloore, "The Implementation of Fuzzy Algorithms for Control", *Proc. IEEE*, 64(4):572-573, 1976.

A correspondence item that shows how fuzzy control algorithms may be implemented as look-up tables.

[91] N.K. Sinha, J.D. Wright, "Application of Fuzzy Control to a Heat Exchanger Process", *Proc. 16th IEEE Conf. on Decision and Control*, New Orleans, December, 1977.

A description of the application of a fuzzy algorithm for controlling the temperature and rate of flow of water in a continuously stirred tank heat exchanger. The algorithm is a two-input two-output system.

[92] H. Sira-Ramirez, "Fuzzy State Estimation in Linear Dynamic Systems", *Proc. 19th IEEE Conf. on Decision and Control*, Albuquerque, NM, December, 1980.

The problem of state estimation in linear dynamic systems with fuzzy initial states and poorly defined inputs is considered. Recursive state estimation formulae, in the spirit of the Kalman filter, are developed to obtain a fuzzy set of possible states.

[93] W.E. Stein, "Optimal Stopping in a Fuzzy Environment", *Fuzzy Sets and Systems*, 3(3):253-259, 1980.

This paper considers the problem of optimally controlling and stopping either a deterministic or a stochastic system in a fuzzy environment. Dynamic programming is applied to yield an approach that corrects some earlier results of other authors.

[94] M. Sugeno, T. Takagi, "Multi-Dimensional Fuzzy Reasoning", *Fuzzy Sets and Systems*, 9(3):313-326, 1983.

This paper discusses a method for describing fuzzy reasoning when the antecedents of rules are functions of more then one variable. The paper argues that treating them as conjunctions or as a collection of one-dimensional implications is incorrect. Several simple examples are given.

[95] M. Sugeno, T. Terano, " Analytical Representation of Fuzzy Systems", In: *Fuzzy Automata and Decision Processes*, Gupta M.M., Saridis G.N. and Gaines B.R. (eds), North Holland, 1977.

This paper contains an attempt to analyze deterministic systems with fuzzy disturbances. The approach is based on the use of fuzzy integrals and a simple example of the derivation of a fuzzy system is shown in the final part of the paper.

[96] T. Takagi, M. Sugeno, "Derivation of Fuzzy Control Rules from Human Operator's Control Actions", *Proc. IFAC Symp. on Fuzzy Information, Knowledge Representation and Decision Analysis*, Marseille, July 1983.

This paper considers the problems of deriving fuzzy control rules by observing and recording the actions of a human process operator. Several procedures are considered and the techniques illustrated by means of a wastewater treatment process example.

[97] R.M. Tong, "An Assessment of a Fuzzy Control Algorithm for a Nonlinear Multivariable System", In: *Discrete Systems and Fuzzy Reasoning*, Mamdani E.H. and Gaines B.R. (eds), Tech. Report EES-MMS-DSFR-76, Queen Mary College, London, January 1976.

A detailed description of results obtained with a controller for a well known two-input two-output system. The fuzzy controller developed compares favorably with more conventional designs and appears more robust with respect to parameter variations.

[98] R.M. Tong, *Some Problems with the Design and Implementation of Fuzzy Controllers.* Tech. Report CUED/F-CAMS/TR127, Cambridge Univ. Eng. Dept., 1976.

A discussion of some practical issues in controller design.

[99] R.M. Tong, "Analysis of Fuzzy Control Algorithms Using the Relation Matrix", *Int. J. Man-Machine Studies*, 8(6):679-686, 1976.

The beginnings of an attempt to develop an analysis procedure that views controllers as relation matrices.

[100] R.M. Tong, "A Control Engineering Review of Fuzzy Systems", *Automatica*, 13(6):559-569, 1977.

A detailed review of fuzzy control and fuzzy systems. Compares various approaches to design and offers some comments on open research questions and directions for future developments.

[101] R.M. Tong, "Analysis and Control of Fuzzy Systems Using Finite Discrete Relations", *Int. J. Control*, 27(3):431-440, 1978.

A discussion of the fuzzy control problem in the context of a finite discrete relations. Describes some definitions of stability and possible control problems formulations.

[102] R.M. Tong, "Synthesis of Fuzzy Models for Industrial Processes", *Int. J. General Systems*, 4:143-162, 1978.

An attempt to define the concept of a fuzzy model and how it might be constructed from data. Suggests some measures of model adequacy and presents some examples from real data.

[103] R.M. Tong, "The Construction and Evaluation of Fuzzy Models", In: *Advances in Fuzzy Set Theory and Applications*, Gupta M.M., Ragade R.K. and Yager R.R. (eds), North Holland, 1979.

Further discussion of the fuzzy modelling problem, together with some general considerations of performance measurement in fuzzy systems.

[104] R.M. Tong, "The Evaluation of Fuzzy Models Derived from Experimental Data", *Fuzzy Sets and Systems*, 4:1-12, 1980.

A concise discussion of the modelling problem, focussing on some the practical issues of model construction.

[105] R.M. Tong, "Some Properties of Fuzzy Feedback Systems", *IEEE Trans. Systems, Man and Cybern.*, SMC-10(6):327-330, 1980.

A detailed analysis of fuzzy systems from a relational perspective. Gives some conditions for being able to derive the solution of various feedback systems.

[106] R.M. Tong, M.B. Beck, A. Latten, "Fuzzy Control of the Activated Sludge Wastewater Treatment Process", *Automatica*, 16(6):695-702, 1980.

A detailed example of the application of the fuzzy rule-based paradigm to a complex industrial process.

[107] Y. Tsukamoto, T. Takagi, M. Sugeno, "Fuzzification of L-Aleph-1 and its Application to Control", *Proc. IEEE Int. Conf. Cybernetics and Society*, Tokyo, November, 1978.

This paper proposes a fuzzy controller based on a fuzzy logic derived from Lukasiewicz's infinite-valued logic. The applicability of the proposed method is discussed by considering a two variable control system with a compensation for time delay.

[108] M. Tsukiyama, T. Fukuda, "An Application of Knowledge Base to Control Systems", *Proc. IEEE. Int. Conf. on Cybernetics and Society*, Atlanta, October 1981.

An interesting non-fuzzy rule-based approach to control. The research is motivated by the same philosophical background as the fuzzy controller work, but achieves its goals with control rules that are deterministic.

[109] I.B. Turksen, *Analysis of Fuzzy Logic Controller*, Tech. Report 78-012, Dept. Industrial Eng., Univ. Toronto, 1978.

[110] I.G. Umbers, P.J. King, "An Analysis of Human-Decision Making in Cement Kiln Control and the Implication for Automation", *Int. J. Man-Machine Studies*, 12(1):11-23, 1980.

This paper describes the investigation of the control skills of cement kiln operators with a view to modelling them by a fuzzy algorithm. The implementation of the prescriptive rules as a fuzzy controller is discussed.

[111] M. Uragami, M. Mizumoto, K. Tanaka, "Fuzzy robot Controls", *J. Cybernetics*, 6(1-2):39-64, 1976.

This paper considers the application of fuzzy programs to the control of an inchworm robot. In the execution of a fuzzy program, fuzzy instructions are translated into machine instructions by the use of the authors own max-method and backtracking. An example of the robot searching for a goal node in a graph is given to illustrate the techniques.

[112] P.P. Wang, M. Togai, "Fuzzy Sensitivity Analysis and Synthesis Technique", In: *Fuzzy Set and Possibility Theory*, Yager R.R. (ed), Pergammon Press, 1982.

A discussion of the problem of sensitivity in dynamic systems. The authors describe the use of fuzzy sets to model system parameters, transfer functions and outputs. The result is a more systematic technique which can solve some problems on which conventional methods fail.

[113] J.J. Weiss, M.L. Donnell, "A General Purpose Policy Capturing Device using Fuzzy Production Rules", In: *Advances in Fuzzy Set Theory and Applications*, Gupta M.M., Ragade R.K. and Yager R.R. (eds), North-Holland, 1979.

A canonical form for fuzzy control rule is described. A simple example of a temperature controller is discussed to illustrate the approach.

[114] F. Wenstop, "Deductive Verbal Models of Organizations", *Int. J. Man-Machine Studies*, 8:293-311, 1976.

This paper pursues the idea that complex systems might be modelled more significantly by using verbal procedures. The author shows how fuzzy sets can be used to make this concept operational and gives some details of a computer system that can be used to perform linguistic simulations.

[115] D. Willaeys, "Optimal Control of Fuzzy Systems", *Proc. Int. Congress on Applied Systems Research and Cybernetics*, Acapulco, December 1980.

This paper contains a mathematical discussion of the problem of obtaining an optimal controller design from a fuzzy description of the system.

[116] D. Willaeys, "Decision Making with Possibilistic Dynamic Systems", *Proc. 20th IEEE Conf. on Decision and Control*, San Diego, December 1981.

Another paper by the author in the same vein as the reference above. The analysis uses the concepts of decision function as introduced by Bellman and Zadeh.

[117] D. Willaeys, P. Mangin, N. Malvache, "Use of Fuzzy Sets for Systems Modelizing and Control: Application to the Speed Control of a Strongly Perturbed Motor", *Proc. 5th IFAC/IFIP Int. Conf. on Digital Computer Applications to Process Control*, The Hague, 1977.

A presentation of the experimental framework and results of applying the fuzzy controller paradigm to a laboratory scale motor system.

[118] D. Willaeys, N. Malvache, P. Hammad, "Utilization of Fuzzy Sets for System Modelling and Control", *Proc. 16th IEEE Conf. Decision and Control*, New Orleans, December, 1977.

The paper contains a brief review of techniques for building models of fuzzy systems for use in controller design. It also describes a new method for deriving a controller given such a model.

[119] D. Willaeys, N. Malvache, "Use of Fuzzy Model for Process Control", *Proc. IEEE Int. Conf. on Cybernetics and Society*, Tokyo, November, 1978.

Similar in content to the previous entry.

[120] S. Yasunobu, S. Miyamoto, H. Ihara, "Fuzzy Control for Automatic Train Operation System", *Proc. 4th IFAC/IFIP/IFORS Int. Congress on Control in Transportation Systems*, Baden-Baden, April 1983.

A good example of the use of fuzzy rule-based techniques to a real control problem. The controller developed by the authors appears to give better more robust performance than conventional controllers for the same problem.

[121] M. Yonekara, "The Application of Fuzzy Sets Theory to the Temperature of Box Annealing Furnaces Using Simulation Techniques", *Proc. 8th IFAC World Congress*, Tokyo, 1981.

[122] L.A. Zadeh, "Towards a Theory of Fuzzy Systems", In: *Aspects of Network and Systems Theory*, Kalman R.E. and DeClaris N. (eds), Holt, Rinehart, Winston, 1971.

A detailed description of a formalism for the analysis of fuzzy systems. The paper contains many ideas that have been employed by subsequent authors.

[123] L.A. Zadeh, "A Rationale for Fuzzy Control", *Trans. ASME.*, 94(Ser. G), J. Dynamic Sys., Measurement and Contr., 3-4, 1974.

A discussion of the "over-mathematization" of control theory and an outline of how fuzzy sets might be used to make complex real world control problems tractable.